DIE CASTING
ENGINEERING

A Hydraulic, Thermal, and Mechanical Process

DIE CASTING
ENGINEERING

A Hydraulic, Thermal, and Mechanical Process

Bill Andresen

CRC Press
Taylor & Francis Group
Boca Raton London New York

CRC Press is an imprint of the
Taylor & Francis Group, an **informa** business

CRC Press
Taylor & Francis Group
6000 Broken Sound Parkway NW, Suite 300
Boca Raton, FL 33487-2742

First issued in paperback 2019

© 2005 by Taylor & Francis Group, LLC
CRC Press is an imprint of Taylor & Francis Group, an Informa business

No claim to original U.S. Government works

ISBN-13: 978-0-8247-5935-3 (hbk)
ISBN-13: 978-0-367-39356-4 (pbk)

Visit the Taylor & Francis Web site at
http://www.taylorandfrancis.com

and the CRC Press Web site at
http://www.crcpress.com

Preface

This is a broad technical presentation for participants in the
die casting process. It is intended that the material presented
will help to reduce manufacturing costs, increase productivity, and enhance quality through failure avoidance. While
the scope is broad and covers the many facets of casting, the
focus is on function, problem identication and solution, and
strategic logic.

All casting processing are a function of velocity and pressure. Die casting is at the high level of both, a fact that
presents unique challenges discussed in this book.

Die casting is the shortest route between raw material
and near net shape.

Acknowledgment

To Barb, who totally supports a hectic career in die casting,
which is so enjoyable that it can hardly be considered work.

Bill Andresen

About the Author

BILL ANDRESEN is the President of Hi Tech International, Inc., Holland, Michigan. An international technical and management consultant, he has a wide breadth of experience in the field ranging from hands-on engineering to the management of manufacturing facilities and service as Technical Director of the American Trade Association.

Bill Andresen has held the positions of Production Control Manager, Chief Engineer, Plant Manager, and Manufacturing Vice President, and was a member of the board of directors and executive committee for *Du-Wel Products*. He was the founding manager of the aluminum plant in Dowagiac, Michigan that historically returned one-half of corporate earnings on one-third of the sales. He also served as the Executive Vice President of *Viking Die Casting Corporation,* where he introduced new technologies that grew productivity and sales. Bill became a disciple of the world class developments by CSIRO in Australia for die casting technology. As a result of the gap between this and actual die casting practice, he formed *Hi-Tech International, Inc.* in 1989. This firm offers measuring, analyzing, designing, and verifying for both new and existing projects to establish true value streams. Quality and productivity enhancement, mechanical die design, flow analysis, and thermal management are available to clients worldwide that are engaged in high pressure die casting.

Mr. Andresen has been a longstanding and active member of the North American Die Casting Association. He served as Technical Director for the American Die Casting Institute where his responsibilities included the management of technical research and worldwide interaction with die casting firms. Considerable public speaking and many published articles on die casting have also left his mark on the industry. He has represented the industry in negotiating more reasonable environmental regulations with various agencies of the United States government. He has served as chairman of the Die Casting Research Foundation and received the Nysellius Award, the highest recognition by the industry for technical contributions.

Mr. Andresen has taught die casting courses at the University of Wisconsin, Western Michigan University, Southwestern Michigan College, and NADCA, as well as teaching the die casting process to individual companies.

Mr. Andresen graduated from Purdue University in West Lafayette, Indiana.

Contents

Introduction

WHAT IS DIE CASTING?

Die casting is a manufacturing process for producing accurately dimensioned, sharply defined, smooth or textured surface metal parts. It is accomplished by injecting liquid metal at fast velocity and under high pressure into reusable steel dies. Compared to other casting processes, die casting is at the top end of both velocity and pressure. The high velocity translates into a very turbulent flow condition. The process is often described as the shortest distance between raw material and the finished product. The term die casting is also used to identify the cast product.

HOW ARE DIE CASTINGS PRODUCED?

First, a steel mold, which is usually called the die and contains the cavities that form the castings, is made into two halves to permit removal of the castings. This die is capable of producing thousands of parts in rapid succession. The die is then mounted securely in a die casting machine with the

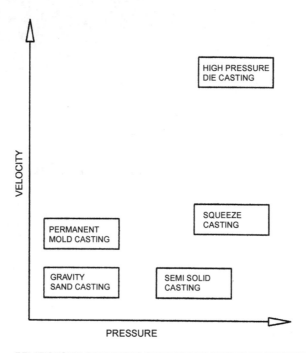

RELATIONSHIP OF VARIOUS CASTING PROCESSES TO PRESSURE AND VELOCITY

Figure 1

individual halves arranged so that one is stationary (cover die) while the other is moveable (ejector die).

The casting cycle starts when the two dies are clamped tightly together by the closing mechanism of the machine. Liquid casting alloy is then injected into the die in an extremely short period of time and at very high pressures, where it solidifies rapidly. The die halves are then drawn apart when the machine opens, and the shot which includes the castings is ejected.

Die casting dies range from simple to complex and have moveable slides and cores as determined by the configuration of the part. They consist of mechanical features; a metal flow system called runners, gates and vents; and a thermal system because the die also acts as a heat exchanger.

The complete cycle of the die casting process is by far the fastest method known for producing precise nonferrous metal castings. This is in marked contrast to sand casting which requires a new sand mold for each casting cycle. While the permanent mold process uses steel molds instead of sand, it is considerably slower and, like sand casting, not as precise as die casting.

BASIC TO THE PROCESS

The die casting process is fundamentally simple but it is complicated by a massive array of ancillary equipment and details. There are only three basic factors (see below) that affect the final product that results from the rapid conversion of metal in the ingot form to a net shape.

Some assumptions are usually made when dealing with die casting that help to visualize the logical chain of events that occur during each cycle. These assumptions are:

- Since the casting alloy is injected into the die cavity at a superheated temperature, it behaves like a hydraulic fluid during the very brief period of cavity fill.
- The metal travels in a straight line until it meets an obstruction and then the stream splashes and breaks up into turbulent eddies. During cavity fill, it follows the path of least resistance.
- Die casting is a turbulent process since liquid casting alloy travels through the system at extremely high rates of speed.

The three fundamental factors are:

- The thermal behavior of the casting alloy that can be quantified by the thermal constants.
- The shot end of the casting machine and the shot sleeve or goose neck that provide the liquid metal required to fill the die cavity.
- The shape of the part that defines the flow path of the liquid metal as it travels through the cavity. The

surface area to volume ratios and the distance that the metal must travel are important mathematical characteristics of each net shape.

This text will attempt to present the details of die casting process in a logical manner. It is definitely predictable and controllable.

AUTHOR'S NOTE

The data presented in this text have been collected by the author from experience and many sources believed to be reliable. However, no expressed or implied warranty can be made to its accuracy or completeness. No responsibility or liability is assumed by Hi Tech International, Inc. or the author or the publisher for any loss or damage suffered through reliance on any information presented or included here. The final determination of the suitability of any information for the use contemplated for a given application remains the sole responsibility of the user.

1

Terms Used in Die Casting

Many texts place this topic at the end or in a separate appendix, but it is addressed here at the beginning so that everyone referencing the subject of die casting may speak the same language. Clear communication is sometimes difficult, yet it is critical to successful die casting.

This is a partial list of the more commonly used terms and is not intended as a comprehensive, totally inclusive glossary. It is intended only to help introduce the subject and, as a convenient reference.

Accumulator: A reservoir in the hydraulic system that holds the shot pressure at a constant level and reduces normal fluctuations. This is a cylinder that is usually located at the shot end of the die casting machine.

Aging: A change in the metallurgical structure, physical properties, and dimensions of an alloy that takes place over an extended period of time after a part is die cast. Aging time is compressed with heat.

Alloy: A metallic material that consists of two or more chemical elements whose physical properties are normally different than those of the separate ingredients.

Anodizing: A process that utilizes the casting as the anode in an electrolytic cell so that a protective or decorative film can be applied to the surface.

ANSI: American National Standards Institute.

AQL: Acceptable Quality Level as agreed upon between the die caster and customer.

Area (projected): The area of the cavity and metal feed system that is visible when viewing the die at an angle perpendicular to the basic parting plane.

Area (surface): The area of the cavity surface that comes into contact with the casting alloy in both die halves.

ASQC: American Society for Quality Control.

ASTM: American Society for Testing and Materials.

Australian metal feed system: A series of tapered tangential runners that are designed to generate constant gate speeds as the casting alloy exits the runner and enters the die cavity. The spurt of energy that occurs at the end of each runner branch is controlled with a shock absorber at this point in the system.

Austenite: A Phase that Iron-carbon steels reach during heat treating that is relatively ductile with a low work hardening rate.

Back scrap: Runners, gates, biscuits, overflows, trimmings, and defective castings that are normally remelted for another try at production.

BHN: A number that quantifies hardness in the Brinell system.

Biscuit: Excess of ladled metal remaining in the shot sleeve of the cold chamber process. It is a part of the cast shot and is ejected from the die with the runner and casting.

Blister: A surface bubble caused by expansion of entrapped gas as a result of excess heat.

Blow holes: Voids or pores which may occur due to entrapped gas or volumetric shrinkage during solidification. This condition is usually evident in heavy sections.

Buff: To smooth a casting surface with a rotating flexible wheel to which fine abrasive particles are applied in liquid suspension, paste, or grease stick form.

CAD: Computer aided design.

Captive: An original equipment manufacturer that produces die castings exclusively for its own use.

CASS test (copper accelerated salt spray): An accelerated corrosion test for electroplated substrates (ASTM 368–68).

Casting alloy: The material from which the die casting is produced.

Casting rate: The average quantity of shots that can be produced from a particular die in one hour of constant running.

Casting/shot ratio: Volume or weight of usable casting product divided by the total volume or weight of metal injected into the die that is expressed as a percentage.

Casting yield: The net number of acceptable castings that are produced from a production run compared to the gross number of shots. It is usually expressed as a percentage. Yield is also sometimes referred to as the ratio of total shot volume to net casting volume expressed as a percentage.

Casting cycle: The total number of events required to produce a high pressure die casting that usually consists of metal injection (including cavity fill) solidification, ejection, and die spray.

Casting drawing: The detailed engineering description of the shape to be cast that defines the size (dimensions), shape, material, and allowable tolerances.

Cathode: The electrode used in electroplating at which metallic ions are discharged, negative ions are formed, or other reducing activities take place.

Cavity: The recess or impression in the die steels in which the casting is formed.

Cavity fill time: The critical time that it takes to fill the die cavity. This time has a profound effect upon the amount of premature solidification that occurs before the cavity is completely filled with metal.

Cavity insert: A die component that forms the shape to be cast.

Cavitation: The generation of cavities in a fluid that occurs when local pressure falls below the vapor pressure of the fluid whenever bubble nuclei are available.

Charpy: An impact test in which the specimen forms a simple beam that is struck by a hammer while supported at points that are 40 mm apart.

Checking: Heat crazing of the surface of the die steel that is manifested in a series of fine cracks caused by extreme thermal fatigue. Corresponding raised veins on the casting surface are formed when this condition occurs.

Chisel gate: A gate shaped like the point of a chisel which is designed to direct a single stream of metal straight into a specific target location within the die cavity.

Chromate: A conversion coating of trivalent and hexa-valent chromium compounds.

Chrome pickle: A chemical treatment for magnesium castings that provides some protection from corrosion or oxidation when a dichromate film of nitric acid is formed.

Clamping capacity: The ability of each tie bar to hold the machine platens and die halves together during the injection of metal under high pressure. Also the number that describes the size of the casting machine.

Clamping force: Actual force applied to a particular die during metal injection. This is less than the machine capacity.

Cooling medium: The liquid—either water, steam, or oil—that is utilized to remove the heat conducted into the die steels by the injection of liquid metal during each casting cycle.

Cold chamber: A die casting process in which the metal injection mechanism is not submerged in liquid metal.

Cold shut: Poor fill or surface finish in a die casting caused by low metal or die temperatures.

Combination die: A die with two or more cavities in which each cavity forms a different shape.

Compressive yield strength: The maximum compressive stress that a die casting can withstand without a pre-determined amount of yield (usually 0.2%).

Constant area sprue: A sprue post that is designed with a gap between the male post and the female sprue bushing that decreases as the diameter increases, so that the theoretical area through which the casting alloy travels is the same or less than the area of the nozzle.

Corrosion: Surface condition caused by exposure to gasses or liquids that attack the base metal. Rust on steel is an example.

Constant acceleration: A condition during which the shot plunger continuously advances at increasing velocity from the static position to the end of the shot cycle. This process is favored by European die casters.

Constant velocity: A condition during which the shot plunger advances at a set velocity until it reaches a predetermined position and then increases in velocity until the end of the shot cycle. This process is favored by North American die casters.

Contraction: The volumetric shrinkage that occurs in metals during solidification.

Core: A casting die component that forms an internal feature that is separate from the die insert. It may be stationary and perpendicular to the parting plane or may be located in another direction to be actuated by a movement each time the die is opened.

Cored hole: Any hole in a die casting that is formed by a core in the die casting die.

Cover die: The stationary die half that is mounted to the platen at the shot end of the die casting machine.

Cover gas: A mixture of gases made up of sulfur hexafloride, carbon dioxide, and air that is used to protect the surface of liquid magnesium by reducing the formation of oxides.

Creep: Plastic deformation of metals (zinc alloys especially) that occurs below the yield strength.

Critical dimension: A dimension that must be held within a specific tolerance limit in order for the part to function within its product application.

Custom: A firm that produces die castings custom designed for the exclusive use of an original equipment manufacturer in their end product.

Damping: Refers to the ability of a casting alloy (magnesium) to resist vibrations that lower noise levels.

Deburr: Removal of sharp edges or fins by manual, mechanical, chemical, or electrical discharge methods.

Dendrite: A crystal with a branching tree like pattern that usually is most evident in cast metals that are slowly cooled through the solidification state.

Deflection: The bending or twisting of a shape that occurs when a load is applied to it. Normally, this term is used to describe elastic strain so that it will return to its original form when the load is removed.

Dichromate: A chemical treatment in which aluminum, magnesium, or zinc castings are boiled in a dichromate solution that produces a protective film to minimize corrosion.

Die: Two metal blocks that incorporate the cavity, metal feed system, and thermal channels into the tool that is used to produce die castings.

Die blow: The distance that the two die halves are forced apart by the injection pressure during cavity fill.

Die casting: A process in which a die casting is formed by a mass of molten metal by forcing a heat flux through a mold onto the liquid mass affecting solidification. The resultant solidification patterns and rates determine whether or not the casting satisfies the customer's requirements.

The processing theory defines a step-by-step analytical procedure to design the energy exchange functions necessary to make a useful piece part. The results are the specifications for the die design and the process control set points.

Cooling and/or heating channels plus the heat flow paths must be designed to focus the correct amount of energy through the cavity surface to achieve the required heat flux. Hence, the die design is derived from the defined required final condition of the solidification pattern.

The design of the die includes, in the mechanical aspect: material selection, insert seams, and clearance space; in the thermal exchange, location, size, length of the cooling/heating channels, and the flow rate of the medium used; and in the fluid flow arena: the location and size of the gating and venting, as well as configuration of the metal feed system.

This term is also used to define the net shape produced from this process.

Die life: The number of acceptable shots of castings that can be produced from a die casting die before it must be replaced or extensively repaired.

Die lubricant: Liquid formulations applied to the die to facilitate release after the casting is formed and to prevent soldering of the casting to the die surface.

Die temperature: Usually refers to surface temperatures of die components that come into contact with the casting alloy. The temperature through the thickness of a die component is very complicated and when dealing with the metallurgy of the die steels this term also applies to deeper temperatures.

Dimensional stability: Ability of a casting or die component to retain its shape and size over a long period in service. This term is also applied to die materials during heat treatment.

Dog leg: A cam that is designed to move a side core the appropriate distance and at the proper time.

Dowel: A guide pin which assures registry between die components, usually located in opposite die halves.

Draft: The angle given to casting walls, cores, and other parts of the die cavity to permit ejection after the shrinkage that occurs during casting solidification.

Drag: A defect that occurs when the casting alloy adheres to the die steel during ejection and results in undesirable grooves in the casting.

Dross: Metal oxides that form either within or upon the surface of a liquid metal bath.

Eject: To press the solidified casting away from the core in the die casting die.

Ejector pin: A rod which pushes the casting off from cores and out of the die cavity.

Ejector flash: A thin fin of metal that is formed during the cavity filling between the ejector pin and the mating hole.

Ejector plate: A plate to which ejector pins are attached that activates them.

Electrolyte: An environment, usually liquid, that conducts electricity accompanied by chemical decomposition that defines the incidence of corrosion.

Electroplate: Electro-deposition of a metallic coating to a substrate (die casting) to improve surface properties.

Elongation: The amount of permanent extension in the locale of the fracture in a tensile test expressed as a percentage of the original gage length.

Erosion: Describes the damage to the die surface that occurs when a high velocity metal stream washes away some of the original die material.

Eutectic: The lowest melting point of a metal in an alloy system.

Fatigue: A series of fluctuating stresses and strains less than the tensile strength of the material that lead to fracture when repeated. In die casting, especially when aluminum alloys are involved, the large thermal gradient that occurs during each casting cycle is the mechanism that initiates fatigue.

Family die: A die that produces more than one distinct shape.

Fan gate: A style of gate that is deigned to fan the metal stream out so that the fill pattern becomes wider as the liquid metal progresses into the cavity.

Feed: A term that applies to the delivery of liquid metal to the die cavity. Also, it refers to packing extra metal into the cavity during intensification to compensate for volumetric shrinkage during solidification.

Fillet: Curved junction of two planes that would meet at a sharp angle without it.

Fill pattern: The configuration of the streams of metal within the die cavity that occur during cavity fill.

Finish: The degree of smoothness of a surface of the die cavity or the casting produced from it. It is quantified by the grit size used in the final polishing.

Finite difference analysis: A computer program that utilizes a three-dimensional model to simulate flow patterns within a shape so that they may be analyzed. The model is meshed into thousands of elements that calculate the differences of conditions between adjacent elements.

Finite element analysis: A computer program that utilizes a three-dimensional CAD model to simulate flow patterns within a shape so that they may be analyzed. The model is meshed into many separate and finite elements that can be studied more easily than the whole shape.

Fit: The precision of the clearance or interference that defines the gap between two mating parts.

Fixture: A device that holds a die cast near net shape in a fixed position while a secondary operation is performed on it to convert it to a net shape.

Flash: A thin fin of metal which occurs at die partings, vents, and around moving cores. This objectional metal is due to working and operating clearances in the die. Also—a verb used to describe the condition that exists when the die halves are not held completely closed.

Flow line: Surface marks on a die casting that trace the metal flow pattern.

Flow rate: The quantity of fluid per unit of time that flows through a specific conduit area. In die casting, this can refer to liquid metal, hydraulic fluid, water, etc.

Fluidity: A condition that defines the ease that a liquid metal will travel through a conduit, at a given temperature, before it solidifies.

Flux: A compound in powder form that is applied to minimize oxide formation upon the surface of a liquid metal bath.

Fracture toughness: The ability of a tool steel to withstand the constant expansion and contraction that occurs in each casting cycle.

Gage: A device that compares a cast or machined dimension or relationship to a specified limit.

Galling: Sliding friction that tears out particles from a metal surface.

Gas: Air or gasses from decomposition of release agents that are vulnerable to becoming encapsulated by super heated liquid metal that is a source of porosity in the casting.

Gate: The orifice through which the casting alloy exits the runner and enters the die cavity. Also—the entire ejected content of the die including casting, gate, runner, biscuit, sprue, overflows, vents, and flash.

Geometric characteristics: Basic elements that form a mathematical language for dimensioning and tolerancing used in form, orientation, profile, eccentricity, and location.

Gooseneck: The main metal pressure component for the hot chamber process that contains the shot chamber and also forms a spout at the other end to funnel the casting alloy into the nozzle. The gooseneck is submerged into the bath of liquid metal supply.

Grain: A description of the crystalline structure of the atomic structure.

Grain structure: The size and shape of the grains in a metal.

Growth: Expansion of a casting (more often zinc) as a result of aging, intergranular corrosion, or both.

Hard spot: A dense inclusion in a casting that is harder than the surrounding matrix.

Hardware finish: A description of a very smooth surface that is free of defects and capable of supporting diffuse and specular reflectance. Very high quality and lustrous finish like powder coating or electroplating.

Heat checking: (*see* Checking)

Heat sink: A massive shape whose volume to surface area ratio is greater than the adjacent casting segment that has a greater capacity to hold heat.

Heater: A recess in the die steels, sometimes also called an overflow, that is connected to the cavity by a thin gate. It acts as a heat sink to retain heat at a specific position in the die to reduce problems caused by low die or metal temperatures.

Also—an electric cartridge-type device to introduce heat into a specific cold position in the die.

Heat transfer coefficient: The rate at which a material will transfer heat per temperature gradient over a specified period of time.

Hot chamber: The die casting process in which the plunger and gooseneck is immersed in liquid metal in the holding furnace.

Hot metal delivery: The practice of transporting metal up to 300 miles, from the smelting supplier to the die casting plant in the super heated liquid state, rather than in solid ingot form. There is an obvious energy saving since the metal needs no further melting, but sophisticated scheduling is necessary to ensure that there is holding furnace capacity to receive it.

Hot short: A term used to describe an alloy that is brittle or lacks strength at elevated temperatures.

Hot crack or tear: A fracture caused by thermal contraction stress that occurs just below the solidifying temperature.

Impact strength: Ability of a component to resist shock as measured by a suitable testing method.

Impression: Cavity in a die casting die.

Also—the mark left by a hit from another hard surface.

In the white: A term used to describe the condition of a casting that has not received any finishing or treatment of any kind beyond gate removal.

Ingot: Casting alloy formed in a convenient shape for storage, shipping, or remelting.

Inject: To force liquid metal into a die.

Insert: A piece of material with better properties than the metal being cast, of hardness, strength, etc., usually ferrous, which is placed in a die cavity before each shot. When liquid casting alloy is cast around it, it is integrated into the part.

Also—a separate component in the die casting die with enhanced qualities of fracture toughness where the die steels "see" the alloy to be cast.

Intensification: A hydraulic process that increases the injection pressure (usually by a factor of 3) upon the metal after the cavity is filled to force or to pack more metal into the cavity to increase casting density.

Intergranular corrosion: An attack on grain boundaries (usually zinc alloys) that results in deep penetration and weakness planes.

Izod: An impact test in which the specimen is clamped at one end and acts as a cantilever when struck by a hammer.

Jewelry finish: The highest quality electroplated surface finish for a die casting.

Leader pin: A pin located in one die half to align it to the opposite half.

Leader bushing: A female bushing that is designed to accept the leader pin located in the opposite die half to align the dies.

Leveling electroplate: Electroplate layer of metal (acid copper is a good example) that generates a surface smoother than the substrate.

Liquid: Reference to the state of the casting alloy. Preferable to the word "molten" since the safety connotation is more positive.

Logo: A symbol that identifies the producer of the die casting, often cast into the surface of the part, with the customers permission.

Lot size: The quantity of parts produced from a single die and machine set up.

Loose piece: A type of core that forms an undercut that is positioned in, but not fastened to a die. It is arranged so that it is ejected with the casting from which it is eventually removed. It is used repeatedly for the same purpose.

Manifold: A system that may be located internally or externally to collect several thermal systems into a single system for quicker connection.

Martensite: The hardened micro structure of die steel in which die casting dies display the best performance.

Metal: The material from which the die casting is produced.

Metal saver: Core used primarily to reduce the volume of metal in a casting and to avoid sections of excessive mass.

Multiple cavity die: A die having more than one duplicate impression.

Molten: Liquid state with reference to casting alloy (not politically correct as it connotates a hostile safety condition).

Moving core assembly: Includes the mechanism of gibs, ways, locking wedges, angled pins, dog leg cams, racks, pinions, and hydraulic cylinders that hold and move cores in a direction other than parallel to the die parting.

NADCA: North American Die Casting Association, a die casting trade association in North America that is the consolidation of the American Die Casting Institute (ADCI) and the Society of Die Casting Engineers (SDCE).

Net shape: Form that is die cast; a more scientific name for a die cast part.

Nitriding: A heat treating process that is intended to improve the fracture toughness of die materials by diffusing nitrogen into the surface.

Nozzle: A tubular fitting which joins the gooseneck in a hot chamber process to the sprue bushing in the cover die.

Operating window: The best combination of process variables that will yield the greatest throughput of high quality castings.

Overflow gate: A passage that connects the cavity to an overflow.

Overflow well: A recess in a die connected to the cavity by a thin gate to assist in venting.

Oxidation: A chemical reaction between an alloy, like magnesium, and oxygen or an oxidizing agent.

Parting line: The mating surface, sometimes called the parting plane, between the cover and ejector die halves.

Also—the mark or raised line on the casting that is formed by the interface between the die halves.

Parting line step: A region of the parting plane where the level abruptly changes to accommodate a detail of the part to be cast.

Pitting: Small depressions in the cavity die steel that produce small mating bumps on the casting.

Plastic deformation: Permanent bending or twisting that occurs when a load is applied that exceeds the elastic limit of the material. In die casting, this term usually refers to a casting that is ejected before it attains its full strength.

Platen: Thick plates in a die casting machine or trim press. The die is mounted to two of the platens and the other supports the closing mechanism and tie bars.

Plunger: Ram or piston that forces liquid metal into the metal feed system.

Plunger tip: The feed system component that applies pressure to the casting alloy and injects it into the shot sleeve. Port-opening in the gooseneck (hot chamber process) through which liquid metal enters the injection chamber.

Poka yoke: (A Japanese word for mistake proofing, it is pronounced POH-kahYOH-kay) A detail, device, or mechanism that either prevents a mistake from being made or makes the mistake obvious at a glance.

Polish: To smooth down roughness of a parting line or casting surface with a high speed endless belt coated with abrasive material.

Port: Hole between the metal bath and the shot cylinder through which liquid metal enters a hot chamber metal feed system.

Pouring hole: Opening in the top of the shot sleeve into which liquid metal is poured.

Porosity: Voids or pores in a casting that are caused by entrapped air (gas porosity) or volumetric shrinkage during cavity fill (shrinkage porosity).

Preheat: The practice of heating a die casting die to at least 200°F above ambient temperature to minimize the thermal shock from the first few shots in a production run.

Primary alloy: An alloy whose main element comes directly from the natural ore.

Process Control: Control of the process variables within an acceptable range so that high quality castings are produced by the manufacturing process.

Process monitor: A measurement of actual process variables that may be compared to theoretical conditions.

Pressure tight: A casting requirement for internal integrity in which fluid or air, under a specified pressure, will not pass through the casting wall.

Quench: The cooling in a bath, usually water, of a casting from ejection temperature (400–600°F) to ambient room temperature (80–100°F).

Also—used with relation to heat treating of die materials in a vacuum or salt bath, when dropping from austenitizing (1850°F) to tempering (1200°F) temperature.

Quick die change: A procedure of standardization and efficiency to reduce the set-up time of the die casting die.

Radiology: A picture, such as an x-ray, that reveals flaws in the internal integrity of a particular casting.

Rapid prototyping: Inexpensive, accurate model of a proposed part design produced more quickly than by traditional methods.

Refine: The removal of magnesium oxide and other nonmetallic impurities from magnesium with flux that preferentially wets them so they are carried to the bottom of the melt as sludge.

Release agent: A liquid that is usually sprayed onto the die surface to keep the casting from adhering to it. The agent is applied, mixed with water in a ratio of approximately 60 parts of water to 1 part of the agent. The water evaporates from the die surface prior to injection of the casting alloy for the next shot.

Refractory: A material that is not damaged by heating to high temperatures.

Remelt: Process of melting back scrap in a break down furnace so that the liquid metal may be reintroduced into production.

Retainer: The die component that contains the cavity inserts in both halves of the die.

Rib: A wall perpendicular to another wall to provide strength or support. In die casting, ribs also are used to feed liquid metal within the cavity during cavity fill. They are also used to minimize twisting and bending due to uneven shrinkage.

Runner: This conduit is the main part of the metal feed system that transfers the casting alloy from the biscuit (cold chamber) or the sprue (hot chamber) to the gate.

Runner sprue: A runner that is machined into the side of a sprue post. The hot chamber post is smaller in diameter to reflect the small diameter of the nozzle and the cold chamber post is larger since the shot sleeve is much larger in diameter.

Satin finish: A surface finish that presents a diffuse reflector that is lustrous but not bright or smooth. Such a finish sometimes can cover surface defects in the casting.

Scale: Usually a combination of the oxide of the casting alloy and the release agent that builds up during the operation of the die.

Secondary alloy: An alloy that consists of a central element that is resmelted from scrap materials. Most aluminum die castings are produced from secondary alloys, while zinc and magnesium castings are made from primary alloys.

Segregation: Erratic distribution of alloying elements, impurities, or microstructure in a bath of liquid metal.

Shot: That part of the casting cycle that injects liquid metal into the die cavity.

Also—the entire ejected content of the die, including casting, gate, runner, biscuit or sprue, and flash.

Shot peen: A practice that produces a compressive stress on the die surface with a high velocity stream of metal shot or glass beads to close small shallow die checks and increase die life.

Shot size: The capacity of a machine and shot sleeve to provide liquid metal to a die expressed by weight or volume.

Also—the volume or weight of a particular shot that includes the metal feed system, overflows, and the casting.

Shrink mark: A depression on the casting surface opposite a section that is more massive than adjacent walls that is caused by uneven cooling.

Shrink factor: Consideration to recognize the different volumetric shrinkage of the various casting alloys by designing the cavity dimensions over those specified by the part design. It is expressed in terms of linear shrinkage times the nominal dimension. Normally, 0.006 inch per inch is used for aluminum and 0.008 inch per inch is used for zinc.

Shrinkage: Volumetric reduction that accompanies the transition of the casting alloy from the liquid to solid state.

Shot: Synonym for a die casting production cycle.

Also—a term used to describe the total volume of metal produced from the casting die including runners, gates, biscuit, overflows, and usable castings.

Shot sleeve: The steel tube, in the cold chamber process, that holds the casting alloy and through which the plunger tip moves the metal into the metal feed system and the cavity.

Shrink: A mark or depression that sometimes occurs on the surface of a casting opposite a massive section such as a rib, because the mass cools more slowly than the adjacent areas.

Also—to reduce in volume.

Shut off: The space on the parting plane of a die that provides an unrestricted area to apply the clamping force of the machine to seal off flash generation.

Shut height: The total dimension of a die from the back of the cover die to the back of the ejector rails that determines the die opening between platens.

Skin: Surface metal on a die casting with a depth of approximately 0.015 inch that displays a fine dense grain structure and is free of porosity.

Slide: A component of the die that is arranged to move parallel or at least not perpendicular to the die parting. The inboard end forms a portion of the die cavity that involves one or more undercuts.

Solidus: A line on a phase diagram that represents temperatures at which freezing ends on cooling, or melting on heating.

Soldering: Adherence of the casting alloy to portions of the die that are too hot.

Sow: Large solid block of aluminum casting alloy that weighs 2,000 pounds.

SPC: Statistical Process Control that monitors deviations in the process variables from the operating window.

SQC: Statistical Quality Control.

Split gate: A gate of castings having the sprue or plunger axis in the die parting.

Sprue: The conical passage between the nozzle or biscuit and the runner.

Sprue post: A tapered male core that projects into the sprue bushing to deflect metal into the runner system.

Sprue bushing: The female insert in the cover die to contain the casting alloy as it travels into the runner system.

Stake: A cold forming process to bend tabs and studs to assemble zinc castings (usually) for assembly onto mating parts.

Steel safe: A strategy used by metal cutters (tool makers) when close tolerances are involved, in which exterior surfaces of the cavity are intentionally machined slightly undersize and interior surfaces oversize. Thus, any dimensional modifications can be made by removing instead of adding die material.

Stereo lithography: A method of rapid prototyping that utilizes three-dimensional CAD (computer aided design) data to form a series of thin slices with a laser generated ultraviolet light beam that traces each layer onto the surface of a vat of liquid polymer. Thus, each layer is formed and hardened until the prototype is completed.

Stress: Force applied to a section.

Strain: The change in shape that occurs when stress is applied beyond the elastic limit of the material. The stress/strain relationship is a characteristic of the particular section.

Substrate: Parent metal onto which coatings are deposited.

Sulfur hexafloride (SF6): A gas mixed in low concentration ($< 1\%$) with carbon dioxide and air that provides a protective atmosphere over the surface of liquid magnesium to minimize burning and oxidation.

Surface treatment: Modification of a surface. This can apply to either castings or die materials.

Thermal system: A series of channels within a die that carry the cooling medium to extract heat conducted into the die by the casting alloy that is above the liquidus temperature during each casting cycle.

Tie bar: Usually, but not always, there are four bars that are fastened to the two stationary platens of the casting machine. These bars stretch during each casting cycle to provide a locking force to hold the dies shut when high pressures are applied to the metal.

TIR: Total indicator reading.

Toggle: The linkage employed to mechanically multiply the force of the clamping system of the die casting machine when the platens are closed.

Tolerance: A specific acceptable range. This term can be applied to dimensions, temperatures, metallurgical elements, etc.

Toughness: The physical property of a material that allows it to bend or stretch without breaking.

Trim die: A die for punching or shearing the flash from the die casting.

Trim press: A mechanical or hydraulic power press used to trim the flash, overflows, and runner from the cast shape with a trim die.

Tumble: A process to remove rough edges from die castings that utilizes a rotating barrel or vibrating hopper filled with polishing media in addition to the castings.

Twinning: A mechanism in which atoms move between planes of a lattice structure to improve ductility.

Unit die: A die designed to accommodate otherwise unrelated dies in a common holder for more economical production.

Undercut: Recess or cored hole positioned perpendicular to the die parting that prevents ejection.

Vacuum: A mechanical system that draws a partial vacuum within the cavity prior to, or in some cases during, cavity fill to assist in evacuating the cavity.

Vena contracta: A scientific phenomenon that occurs when the direction of a liquid stream is changed (from horizontal to vertical). The stream reduces in cross-sectional area and, in so doing, the speed is increased. After the directional change has been accomplished, the area of the stream increases to normal and thus the speed then also reduces. This is one cause of air entrapment and should be minimized where possible.

Vent: A shallow passage off of the die cavity at the last place to receive liquid metal that allows air from the metal feed system to escape as the cavity is filled.

Void: A large pore within the wall of a casting usually caused by entrapped gas or premature solidification.

Wire brush: A practice of deburring, edge blending, and surface finishing by contacting the work surface with a rotating wire brush.

Yield strength: The stress at which a material exhibits a specified limiting permanent strain or deviation of more than 0.2% from the specified relationship of stress to strain.

ZA: A commercial designation for three high (8–12–27%) aluminum content zinc alloys that display extremely good resistance to abrasion and have high tensile strengths.

Zamak: An acronym for zinc, aluminum, magnesium, and copper that designates zinc casting alloy nos. 2, 3, 5, and 7.

2

Product Design

Almost any net shape can be die cast, provided that the size, including volume, is within the capacity range of commercially available machines and liquid metal delivery systems. However, if the commercial and technical advantages of the process are to be realized, each shape must be intelligently designed or, as is sometimes the case, redesigned.

Many die castings are redesigned from other manufacturing methods so that a net or near net shape can be produced in milliseconds. If appropriate changes are not made, strength could be impaired and complicated manufacturing challenges may result in unreasonably excessive costs. Informed die casters understand that economy is probably the main attraction for designers to choose the die casting option when metal components are required.

The degree of difficulty of the net shape of a die casting is an issue that has not been studied seriously by very many die casters. It should be quantified because it affects cost and manufacturing feasibility. Casting cycle time is vulnerable to complexity, but tool cost and die life are also involved. Details of shape are quantitatively described throughout this chapter. Tooling and processing complexity are addressed by El-Mehalani and Miller in their paper "On Manufacturing Complexity of Die Cast Components," via a combination of empirical experience contributed by 15 die casting firms and by mathematical quantification. A coding system is used to calculate the economic effect of specific details based upon the complexity of the geometric shape (El-Mehalani and Miller). An attempt is made here to expose the reader to such a strategy. It should be noted, however, that it is very difficult to generalize individual details because the variety of die cast shapes is infinite. The number of evaluations expands exponentially when items like draft and depth are included.

Ribs, cored holes, and bosses are described in Fig. 1 and then quantified in a spread sheet (Table 1).

The cost effect of each of these details upon tool path programming and cutting time is obvious. The super heated liquid metal flows over a rib detail and then backfills it. Cores and bosses, illustrated in Fig. 2, obstruct the flow and may require strategic cooling, which increases cycle time and manufacturing cost. The spread sheet shown in Tables 1 and 2 suggests appropriate multiplication factors from the flat surface benchmark that represents 1.0 degree of difficulty.

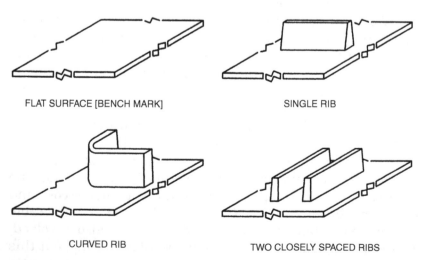

FLAT SURFACE [BENCH MARK]   SINGLE RIB

CURVED RIB   TWO CLOSELY SPACED RIBS

Figure 1

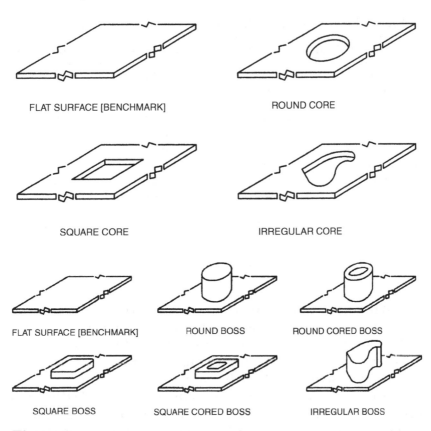

FLAT SURFACE [BENCHMARK]   ROUND CORE

SQUARE CORE   IRREGULAR CORE

FLAT SURFACE [BENCHMARK]   ROUND BOSS   ROUND CORED BOSS

SQUARE BOSS   SQUARE CORED BOSS   IRREGULAR BOSS

Figure 2

Parting line steps and side cores, described in Fig. 3, are more details that complicate both tooling and the die casting process. Table 2 quantifies the cost effects.

Tolerance allowance is necessary in die casting to allow for deviation from shot to shot such as variations in filling and cooling rates. Die wear and deflection can also be expected. Typical benchmark dimensional tolerance is in the range of 0.001–0.010 in., which requires no additional cost factor. Tolerance in the range of 0.001–0.005 in. calls for a degree of complexity of 0.30, and 0.001–0.003 suggests a factor of 0.67. Dimensional tolerance of 0.001–0.002 needs a factor of 1.05.

Table 1 Relative Complexity Due to Individual Details

Description of detail	Tooling factor	Processing factor	Two incidences
Straight rib	0.12	0.13	0.42
Curved rib	0.27	0.27	0.43
Two ribs	0.28	0.28	0.50
Round core	0.10	0.17	0.39
Square core	0.17	0.25	0.44
Irregular core	0.37	0.33	0.43
Round boss	0.12	0.17	0.53
Square boss	0.20	0.26	0.52
Irregular boss	0.31	0.31	0.57
Round cored boss	0.24	0.28	0.58
Square cored boss	0.32	0.35	0.60

Where multiple dimensions are tightly toleranced, an additional factor of 0.09 applies to three occurrences, a factor of 0.18 is necessary for five dimensions, a factor of 0.33 applies to seven incidences, and for 10 tight tolerances, a factor of 0.59 is suggested to reflect the proper cost.

Geometric tolerances are published by NADCA and define both standard specifications and precision values. Standard specifications can be achieved within the commercial cost structure, but precision values are announced at premium cost levels. The calculations here attempt to quantify

Table 2 Relative Complexity Due to Parting Line Steps and Side Cores

Description of detail	Tooling factor	Processing factor	Two incidences
Simple side core	0.38	0.21	0.76
Two side ribs	0.42	0.31	0.79
Side core and rib	0.45	0.38	0.79
Simple parting Line step	0.19	0.14	0.60
Complex parting Line step	0.42	0.42	0.67

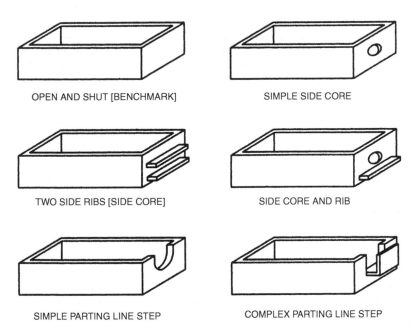

OPEN AND SHUT [BENCHMARK]   SIMPLE SIDE CORE

TWO SIDE RIBS [SIDE CORE]   SIDE CORE AND RIB

SIMPLE PARTING LINE STEP   COMPLEX PARTING LINE STEP

Figure 3

both levels. It has been empirically determined by inquiries to 15 die casting firms, that more than six geometric tolerance specifications describe a die casting that is extremely difficult to produce. It must be noted, however, that one firm reported 10 specifications in successful production. Table 3 lists the degree of complexity for one geometric tolerance for both levels. The benchmark value of 1.0 is a casting net shape with no geometric tolerance.

Die design also affects tooling and processing costs. For this reason, it is most economical to locate details of cavity shape into one die half, if possible. Die halves can shift in the X and Y dimensions and blow apart in the Z dimension. Thus, any dimension or geometry across the parting line is subject to greater deviation from mean than those contained within a single die half. Figure 4 illustrates typical movements.

The top sketch in Fig. 4 represents a shape that can be cast with a flat parting plane between the die halves where

Table 3   Degree of Complexity

Geometric tolerance	Complexity at standard value	Complexity at precision value
Flatness	0.41	0.81
Straightness	0.37	0.74
Roundness	0.28	0.62
Cylindricity	0.28	0.65
Angularity	0.28	0.60
Parallelism	0.32	0.73
Perpendicularity	0.38	0.85
Position	0.27	0.57
Concentricity	0.29	0.67
Runout	0.25	0.62

die blow describes the Z variable and die shift is measured in the X and Y directions. The heavy line depicts the parting line that separates the die steels. This is the simplest die configuration and is used as the benchmark with a degree of difficulty of 1.0.

The lower illustration in Fig. 4 is based upon significant cavity detail formed by both die halves. This additional complexity increases tooling cost by 29% and the process by 13%.

The top sketch in Fig. 5 again depicts a cavity shape that is formed with a flat plane separating the die halves. This, of course, represents a complexity of 1.0. The bottom figure shows a typical parting line step, which increases tooling complexity and cost by 17%. The difference in processing cost between the two is negligible.

Surface finish specification profoundly affects complexity because of the wide variation in metal flows and temperature management encountered in the production of high pressure die castings. NADCA defines the benchmark finish as "as-cast, mechanical grade," with a quantified complexity value of 1.0.

NADCA, "as-cast, paint grade," finishes increase the process complexity by 56%. No die cast tooling is involved. "As cast, high grade" usually relates to polished and buffed

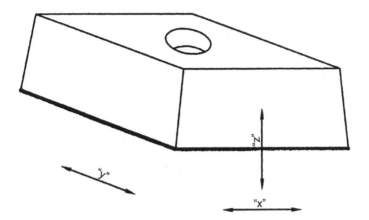

SHAPE CAST IN SINGLE DIE HALF [BENCHMARK]

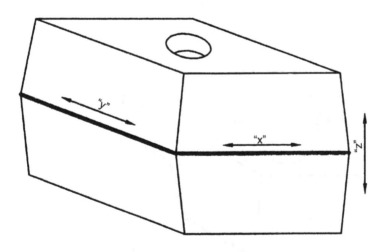

SHAPE CAST IN BOTH DIE HALVES

Figure 4

plated finishes that magnify any surface flaws. This really boosts the processing complexity to 122% of the benchmark!

Wall thickness is usually a compromise between the product designer who desires the thinnest possible wall and processing feasibility where "the thicker the better" is the rule. Actually, given the vicissitudes of the process, it is not

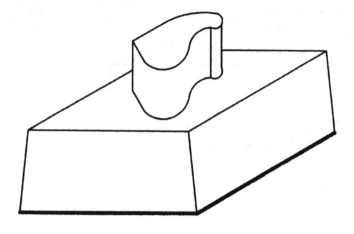

FLAT PARTING PLANE [BENCHMARK]

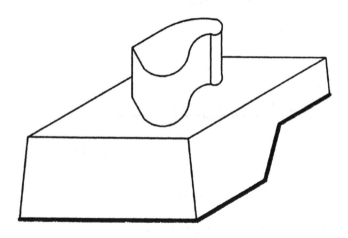

STEPPED PARTING PLANE

Figure 5

wall thickness that determines complexity, it is the surface area to volume ratio (SA/V) that determines the percentage of premature solidification that can be expected during cavity fill. The benchmark is different for each casting alloy; the degrees of difficulty are expressed by Table 4.

Remember that the above data were empirically collected from a number of die casters to demonstrate the fact that quantification of complexity of product design is possible. This approach will eventually replace "seat-of-the-pants" logic when it can be programmed into CAE software for quicker analysis. The balance of this chapter explains the rationale behind simultaneous engineering of product, die casting, tooling, and process.

The most effective strategy for either a new component design or the redesign of an existing part is to relate the performance requirements to the strengths and weaknesses of the die casting process and the range of casting alloys that are available. Other chapters will cover both in detail. Some background is helpful to accomplish this relationship.

The function or performance of the casting must take precedence over any other factors that emerge from the design explorations. Of course, if there is a fit requirement with mating parts, this becomes just as important, but sometimes the periphery of the part merely fits the air.

A combination of knowledge and experience is necessary for proper design of a component to be die cast. The product designer is qualified in the discipline in which the part must perform, but the necessary intimacy with the die casting process is rare.

Table 4 Degrees of Difficulty of Different Surface Area/Volume (SA/V) Ratios

Casting alloy	Bennchmark SA/V ratio	Maximum SA/V ratio	Minimum SA/V ratio	Complexity quantification
Aluminum	10			0.0
		5		0.46
			20	0.85
Magnesium	12.5			0.0
		8		0.46
			25	0.85
Zinc	15			0.00
		9		0.46
			33	0.85

Many times a die caster will be consulted, which brings familiarity with the process to the table. However, most die castings are procured with competitive bidding and it is very possible that the die casting firm that provides design advice will not get the job. This presents an awkward situation.

This book is intended to provide a reference that can sub-stitute for the lack of expertise in the manufacturing process on the part of the product designer. The author has had many years of hands-on experience in multi-plant die casting engi-neering and management. This is combined with research and world-wide consultation to both die casting companies and original equipment manufacturers who are die casting users, which brings creditability to the advice.

A *theoretical model* has been prepared in the flow dia-gram described here that represents the ideal methodology for product design that is compatible with high pressure die casting. For a comprehensive plan like this, feedback is required from analytical metal flow and thermal calculations and/or simulations. This calls for at least a quasi-partnership between the product designer and the die casting engineer. There are many reasons that stand in the way of such a partnership, but to address them would be an inappropriate diversion.

In Fig. 6, note that it is difficult to separate design from modern controls of quality and production. Therefore, a com-mon procedure used to establish the performance of product quality, is described by the acronym APQP, which means Advanced Product Quality Planning, and is suggested to take place during construction of the die. Another procedure, PPAP, which stands for Production Parts Approval Process, is included to validate the process and tool performance within limits established by the product design.

The functional and cosmetic requirements are clearly stated and become the basic guidelines. The net shape is created by computer aided design (CAD) and solid modeling. Computer aided engineering usually includes finite element analysis or finite difference calculations that include elements like necessary fits with mating components and structural analysis. So far, none of this has anything to do with the die

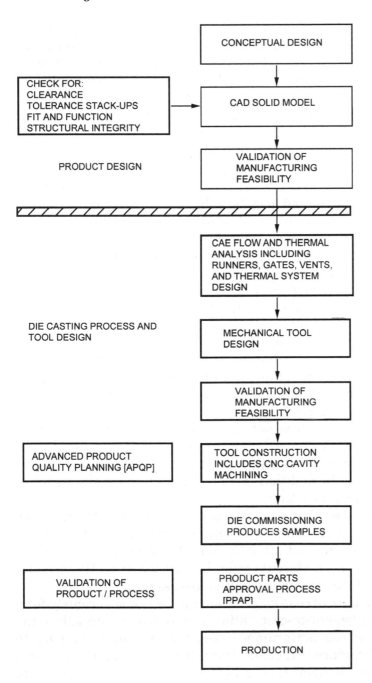

Figure 6

casting process. When a net shape has been designed that functions properly and looks right, it is presented to the die casting engineer so that it may be studied for compatibility with the process that will yield the best possible quality at the lowest piece price and tool cost. A quantitative analysis requires fluid flow and thermal models that address the net shape are provided. These models are then used to develop preliminary windows of opportunity. At the same time, these data are inserted into the design of the actual die casting process. Suggested revisions will emerge from this comprehensive procedure that are fed back into the net shape design for either rejection or acceptance and another iteration is then made until all strategies have been thoroughly considered.

Such a logical procedure will almost certainly yield a die casting die that performs with what is known as "first shot success"—i.e., that produces structurally sound castings that can be dimensionally measured by computerized numerical control (CNC) devices.

After sample approval, optimum die casting production is merely a push of a button away. The alternative is to cut and try seat-of-the-pants methods. Too much guess work and expensive revisions to steel dies, rather than to computer models, is necessary.

Quality, which is not exclusive to die casting, is also a major factor that must be kept in mind. Die castings can be produced with a smooth surface to satisfy the cosmetic requirement. In addition, precise detail is possible, which is unique in metal casting.

Remember that each die casting must be cast into two halves of a permanent steel die and that solid metal cores have to be drawn out after the liquid casting alloy has solidified. Therefore, undercuts must be avoided or costly mechanisms and die construction will be necessary.

The choice of casting alloy should be taken seriously for a maximum benefit-to-cost ratio. It is important to select the material whose performance specifications best compare to the performance expected from the part being designed. The higher temperature alloys like aluminum, brass, and magnesium work best for functional components that require dimensional stability, strength, and wear resistance. The low temperature alloys of zinc are the choice where high quality surface finish is important.

With a few exceptions, almost any shape can be cast in the basic die casting alloys that are commercially available. However, certain alloys offer better value than others with respect for a specific performance requirement.

The different casting alloys behave differently and thus require different technical strategies. The method used for injecting metal into the die cavity is affected, as is the production rate. (This subject is mentioned here only to indicate the several material options available to the designer of die castings. Thermal behavior is discussed in more detail in Chapter 4 on casting metallurgy.)

A few examples here illustrate these comments:

Magnesium alloys offer the best strength-to-weight ratio, but oxidize rapidly and, if this is objectionable, a surface treatment is necessary to overcome it. An ability to dampen sound is an attractive feature of this metal that reduces noise between moving parts. It can be volatile in that it burns extremely bright and hot if in powder or shaving form, and water only exacerbates the fire potential.

Surface finish is more difficult to accomplish with magnesium due to the low liquid density that causes it to freeze more quickly. The cost by volume is artificially held close to that of aluminum for market purposes.

Aluminum alloys are the most popular because they offer a moderate strength-to-weight ratio and are somewhat easier to cast. Unlike magnesium this metal resonates so that castings ring when tapped and are therefore more noisy when performing. In most applications, this feature is not important.

Aluminum is abrasive to die materials so gate speeds have to be throttled back, thus increasing cavity fill time, sometimes compromises quality. This abrasive characteristic limits the die life, which adds frequent replacement of cavity details to the cost structure.

Aluminum is the metal of choice for functional non-cosmetic products and can be cast at slower gate speeds to minimize turbulence within the cavity during fill. It is always cast in cold chamber machines because iron conduits are dissolved rapidly in aluminum.

Zinc alloys are normally used for parts with a cosmetic requirement. This metal is easily plated or coated with paint or powder materials. It is the most fluid of the metal choices and can be more easily cast into thin walls less than 0.06 in. Excellent surface finish is possible when fast gate speeds ensure atomization during cavity fill.

This metal is cast at approximately 400°F lower than the other alloys so is therefore less tolerant of heat.

Zinc alloys are subject to a creep factor that gradually reduces as cast dimensions over a period of time. Heat treating can speed up the creep process, but very tight tolerances are difficult to hold because of this phenomenon.

ZA alloys are a family of zinc to which three different levels of aluminum are alloyed. These alloys provide the best resistance to abrasion and, at the highest level of aluminum, offer tensile strength that compares to mild steel.

Brass alloys are cast at approximately 800°F above aluminum so they are very destructive to die materials and short die life is an expense to be considered. This metal is usually used for plumbing applications.

Lead alloys are cast at approximately 200° below zinc and are subject to rapid freezing during cavity fill. This is the metal of choice for battery connections.

Selection of the parting line is almost akin to establishing the datum line or points. Usually this is considered the responsibility of the die designer, but many times it is too late in the procedure for the die designer to effect the economies that are possible during product design. Good decisions at this stage can pay dividends of faster production rates with peripheral attention to the cost of trimming and finishing.

A flat contour at the parting plane is preferred because both the casting die and trim die will be less expensive and easier to maintain. This does not mean, however that irregular parting lines are impossible or even impractical. They are merely more expensive.

The parting line configuration is important and provides:

- Straight line ejection of the shot after each casting cycle.
- Access for gating and venting.
- Elimination of undercuts.
- Datum for draft calculation and direction.

Even though it is a good idea to keep a design as simple as possible, die casting is a very versatile manufacturing method that can combine separate components of an assembly into a single cast shape. Even when individual components cannot be combined, economies can be achieved, depending on the ductility of the casting alloy, by designing integral fastening details into the net shape that can be riveted, staked, or spun over. Male threads can be cast to avoid a secondary operation, and extraordinary thermal controls can reduce dimensional tolerance requirements to also minimize machining or straightening costs.

Master datums and points usually provide the orientation genesis for all other details of the net shape. Even though these datums may be very familiar to the product designer, sometimes they are not as evident to the tool maker or die caster. It is therefore good to clearly identify them and explain the importance of each.

This information then becomes the basis for locating the shape for subsequent machining and positioning of cavity details in the proper die half relationship for concentricity, alignment, etc. It also defines gage points for quality control of dimensions.

Since die casting is the shortest route from ingot to net shape, considerable production economies are possible. There are, however, several other reasons for the product designer to decide on die casting a proposed shape. Some of these reasons are outlined below:

- Precise detail is possible.
- Dimensional integrity and quality are reasonably predictable.

- Range of casting alloys and physical properties are available.
- Electromagnetic and radio frequency interference shielding for electrical requirements is provided.
- As cast surface finish satisfies many cosmetic requirements.
- Sound damping (magnesium) minimizes noise.
- Pressure tightness is possible.
- Moderate bearing capability and resistance to abrasion is available with certain alloys.

Many times, historical information defines the quality requirements and performance standards of the product to be die cast. In some form or other, most die cast components are derived from earlier versions that were designed to perform the same or a similar function. Quality requirements become field tested standards of performance when the product design is evolutionary.

The most tangible and predictable are the mechanical quality requirements that determine dimensional tolerances that may be linear, effect alignment, call for flatness, etc. Many of these dimensions evolve from the historical function of predecessor parts, and just as important, the ability to produce to these criteria can be confirmed by recorded processing data.

Even the cosmetic specifications that define the surface finish can sometimes be predicted since the ability to produce these elements can be weighed against similar parts produced in the past. The combined experience of the die casting firm that must produce the component also has a lot to do with the final result.

The internal integrity is the most difficult quality requirement to deal with, so considerable emphasis in this text is placed upon the quantification and logical calculations that enhance this aspect. Even this feature is evolutionary because the history of controlling the die casting process goes back to the 1920s, when it was called a black art.

Quantitative advances in the high pressure die casting process within the last two decades have made it possible to die cast net shape or near net shape components. Product designers recognize this potential and constantly challenge the industry to meet higher and more cost effective objectives.

Historical assemblies can sometimes be produced as a single product to eliminate the labor cost to assemble single parts together.

A useful reference is the "Product Specification Standards for Die Castings" that is available from the trade association NADCA.

The performance requirements can affect costs as much as the net shape design.

If only historical data are used as a design reference, the casting shape can sometimes be established by the restraints of whatever process was used to produce predecessor components. This is especially true of parts that were previously produced from other foundry procedures that get strength from heavy walls. It is important to understand that a die casting derives much of its strength from relatively thin walls and fine dense grain structure.

When the final function is the engine that drives the design, the historical shape tends to change because of a new and more flexible set of restraints. It is not necessary for a part to demonstrate excessive strength. It is only important that the component be designed to perform adequately to satisfy the functional requirements.

The function also defines the physical properties of the casting alloys that must be compatible. For example, if a high strength-to-weight ratio is critical, the magnesium alloys present the best option. The aluminum alloys also are lighter materials with good dimensional stability and are especially suited for functional performance where it may be possible to achieve an acceptable fit without secondary machining.

What about the operating environment in which the component must function? It is important to analyze several conditions that have to be satisfied by the product design.

Some things to be considered are listed here:

- Structural requirement
- Cosmetic appearance
- Relationship to other parts in the assembly

- Operating temperatures
- Electrical conditions
- Noise requirements
- Abrasion resistance

Almost every original equipment manufacturing discipline uses die castings. Some of the many applications are:

- Agricultural equipment
- Aircraft
- Automobiles
- Building hardware
- Communication applications
- Computer hardware
- Electrical and electronic equipment
- Home appliances
- Industrial applications
- Instrumentation
- Gardening devices
- Office machines and furniture
- Recreational items
- Toys
- Tools

Most die castings are shells that are defined by their walls and the thickness of those walls.

The thickness depends upon several things.

Strength is a prime requirement of functional die castings since most have to carry some load. One would think that heavy walls would provide the most strength but it is important to note that much of the strength of a die casting comes from the skin. It is about 0.015 in. deep with a very fine and dense grain structure and zero porosity. Therefore, considerable thought must be given to the definition of wall thickness.

Stiffness may be a factor in the effective function of the component being designed. Heavy walls usually are not the way to go here. A well-designed rib pattern brings in the skin strength to address this issue more effectively. Sometimes the rib pattern can be designed like the truss of a bridge or roof to accomplish this objective economically. There are product standards that can be helpful but things like rib design and adjacent elements also must be incorporated into this important decision.

Location of adjacent details with respect to the main body of the shape must be considered because they can sometimes provide the structural integrity necessary for required function.

The environment for the liquid metal streams that form as the casting alloy travels from the gate orifice to the extremity of the part is another critical function that the gap between the two sides of the wall performs. This is key to the relationship between part design and the die casting process.

Fluidity of the casting alloy also plays a critical role in designing wall thickness. Zinc is more fluid than aluminum so wall thickness with this material can and should be thinner. There will be more discussion about this, in addition to recommended wall thicknesses for the different casting alloys, later on in the text.

Complexity of the casting shape will strongly influence the decision on wall thickness. Sometimes it contributes to the strength requirement but it is wise to have a constant wall thickness as the degree of complexity increases.

Casting size, of course, determines wall thickness in that castings over 200 in.^{3} in volume should have walls at least 3/16 in. thick. On the other hand, castings 5 in.^{3} in volume can have wall thickness in the range of 1/16 in., depending on the casting alloy to be used.

The objective is to design a section as thin as possible that will provide sufficient strength and stiffness for the component to function as specified. A useful reference is the "Product Specification Standards for Die Castings" that is available from the die casting trade association (NADCA, 1994).

The design configuration has to be compatible with the die casting process if cost effective productivity and quality are to be expected. To define what this means, the part must:

- Consistently and completely fill with metal.
- It should solidify rapidly without defects.

- • The designed shape must eject easily from the permanent steel dies.

The wall thickness should be as uniform as possible and junctions between walls should blend smoothly. Figures 7 and 8 illustrate the complications of inconsistent wall thicknesses.

Remember, during cavity fill, the super heated liquid metal behaves like a hydraulic fluid and will follow the path of least resistance.

Sharp corners should also be avoided because the metal loses fluidity as the temperature drops and rounded or chamfered corners receive the semi-solid metal more easily than sharp or square details. Some stylists will opine that this compromises the appearance; however, many times this can also be accomplished by challenging the gate design.

As with any casting process, die casting needs appropriate draft away from the parting planes. Therefore, many times nominal dimensions are dimensioned plus or minus draft.

Reverse draft or undercuts must be minimized to avoid secondary machining or moving die parts, both of which adversely affect piece or tool costs. Specifics on this subject will follow later in this chapter.

Each year *Die Casting Engineer* magazine publishes the best casting design and is a good place to observe how close to the edge of the cliff a product designer may walk when married to a good die caster.

An extreme example that this writer will not forget involved a zinc casting that required a 0.750 in. square hole 6 in. long with a dimensional tolerance of 0.004 in. from end to end. This was accomplished with a moving core hydraulically pushed through the opening. Needless to say, the productivity was low and the scream of the core broaching its way through sounded like a police siren! The customer got what they wanted and at the right price, and that is what counts.

Heavy masses can cause voids. The rapid solidification that occurs in high pressure die casting results in volumetric shrinkage at high mass locations. The last place to solidify is

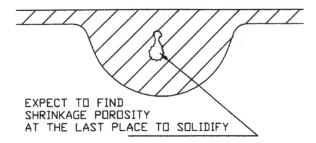

EXPECT TO FIND
SHRINKAGE POROSITY
AT THE LAST PLACE TO SOLIDIFY

Figure 7

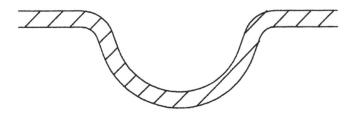

AVOID MASS IN DETAILS FOR INTERNAL INTEGRITY

Figure 8

where shrink porosity can be expected, as seen in Fig. 7. One plan to reduce mass with a metal saver is shown in Fig. 8. Note that shrinkage porosity can be expected in the massive detail.

Alignment of dimensional details is best achieved when each detail is formed by the same die component. Remember that a shift of 0.005 to 0.020 in. can be expected across the parting plane or between a stationary and a moving die member. Several methods to hold die shift to a minimum are covered elsewhere in the text. Concentric diameters can be cast within alignment tolerances as close as modern machining technology can hold them when they are on the same side of the parting line, as described in Fig. 9. The concentricity between diameters on opposite sides is subject to die shift and thermal forces. A rule of thumb is to allow 0.0015 in. total indicator reading (TIR) per inch of dimension plus a factor that is a function of the projected area.

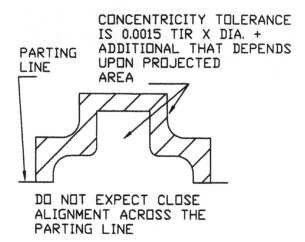

Figure 9

In the case of close dimensional tolerances that are required across the parting plane, it is important to clearly identify the base datum point so that it may be the master locator for machining of dimensions on the other side.

If the objective of effective product design is compatibility between the function of the part and the manufacturing process, a die cast component must exit the die as close to the desired net shape as possible. This is called *near net shape* and secondary manufacturing operations are required to convert it to the net shape defined by the product designer.

Avoid undercuts with a passion because this is the very most economical strategy as far as both the piece part and tooling are concerned. The goal is not to challenge the die designer to execute inventive mechanisms that are possible but not at all economical. Even the simplest core movement in the die casting die costs at least $1,000!

Many times undercuts can be eliminated by a simple change in the casting shape. An example of a design strategy that really amounts to a slight addition to draft angles eliminates the undercut.

Such a potential undercut design consideration is suggested in Fig. 10.

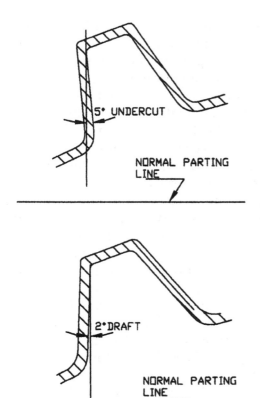

DEALING WITH A POTENTIAL UNDERCUT

Figure 10

Geometric relationships affect freezing schedule is an important concept for product designers to consider so that casting quality and part function can be developed simultaneously. The volume of metal required to fill the part has long been calculated by product designers and cost estimators, so we will start with that.

The ratio of volume to surface area has a major effect upon the degree of difficulty that is required to produce a particular component as a die casting. As this ratio is reduced, the freezing schedule during cavity fill is reduced. This is why wall thickness is a critical design feature. Thinner walls freeze faster than thick walls because the ratio of volume-to-surface-area is greater. Many product designers and die casters as well use a short cut by calculating the average wall thickness to determine a factor to use in an equation to establish the degree of difficulty of a particular near net shape. It is not at all difficult to do the full calculation and the short cut sacrifices too much accuracy to recommend here.

This principle can be explained by comparing the pouring of liquid metal into an ingot, as smelters do, to pouring it upon the floor as sometimes occurs in a die casting plant. In the ingot, the time to solidify is significant. When the same volume of metal is poured on the floor where it can spread out, it solidifies almost instantly. Why is this? The answer, of course, is that the ratio of volume-to-surface-area to is much greater in the ingot than when the metal is poured on the floor.

The opportunities provided in a design for gate locations also have a profound effect upon the productivity so it is logical that a longer periphery or outline of the foot print of the die cavity will provide more opportunities. If liquid metal can be injected into several locations, it is possible for the die caster to devise different fill strategies. This does not mean that more gates make better castings, because just the reverse is true. It simply means that more than one option may be explored.

The distance from the gate to the farthest extremity of the casting also determines the compatibility of the near net shape with the die casting process. Therefore long narrow shapes should be gated across the shortest dimension, all other factors being equal. Usually, if the distance exceeds 8 in. when aluminum is the casting alloy and 4 in. when zinc or magnesium alloys are cast, premature solidification can be expected to occur during cavity fill and cold shut or porosity will result.

Historical data can provide very valuable references with relation to all of these items.

The shape of the component to be produced defines other features that eventually will channel streams of high velocity liquid metal and determine flow paths and direction as liquid metal travels through the cavity. Details are established where air entrapment may cause porosity in pockets that must be back filled, or solder in deep cored holes, etc. Such may be the case even though the shape describes a geometry within the proportional limits discussed above, as shown in Fig. 11.

Conversely, a sharp edge is necessary at the parting line to preclude feather edges in the die steels.

Since many assemblies require a square shape to fit into a radiused corner in a die casting, a depressed radius is suggested in Fig. 12 to allow a radius, that is easier to cast.

While these comments pertain to all casting alloys, they are especially applicable to aluminum alloys that are die cast at higher temperatures that involve higher thermal stresses and are also extremely abrasive.

Each component should be identified, usually by part number, so that it may be visually distinguished from other similar parts in production and assembly. The location of this identification must be clearly specified by the product designer.

The exact specification of the casting alloy is critical since it affects the cost as well as the production strategy.

If the weight or volume of metal required has been calculated, which is normally the case, it should be stated either

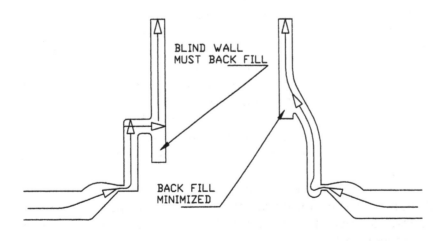

DESIGN OPTION WITH
BLIND WALL

DESIGN OPTION THAT
MINIMIZES BLIND WALL

Figure 11

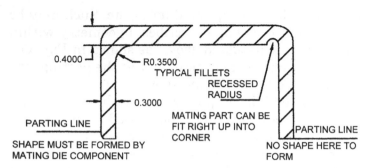

0.4000
R0.3500
TYPICAL FILLETS
RECESSED
RADIUS
0.3000
MATING PART CAN BE
PARTING LINE   FIT RIGHT UP INTO
                            CORNER   PARTING LINE
SHAPE MUST BE FORMED BY   NO SHAPE HERE TO
MATING DIE COMPONENT   FORM

Figure 12

on hard copy or in the CAD files. Even though many three-dimensional CAD systems are capable of this calculation, this ability is not yet universal, and this information is necessary for both cost and process calculations.

Since permanent steel dies are used in the die casting process, draft is necessary in most cases to permit removal of the casting from the dies cavities. This is especially true of internal cores because of the volumetric shrinkage during solidification. Proper draft specification requires the direction of draft to define the maximum material condition.

Datum lines and points normally are the genesis of the casting design and then become the basis for machining location, gage points, and assembly relationships. It is important to locate these datums in the same die half, usually the ejector, to eliminate dimensional discrepancies that occur across parting lines. It is best to cast the datums so that no further consideration is necessary. Along this line, major, minor, and critical dimensional tolerances need to be specified.

Machining stock allowances must be noted and should be designed at the minimum allowed by the dimensional toler-ances, so that the smallest volume of material is removed via machining.

The operating environment must be systematically assessed to determine if the benefits of the high-pressure die casting process will justify the costs.

The temperature of the environment in which the part is expected to operate can be a limiting factor. Aluminum, brass, and magnesium, though considered low temperature alloys, are capable of functioning in higher temperatures than lead and zinc alloys.

It is important to define whether the heat is cyclical or constant. If the maximum temperature occurs only briefly as in some automotive power train components, an aluminum or magnesium alloy will perform without incident. On the other hand, if temperatures in excess of the solidus will be constantly experienced, ferrous alloys must be used.

Both internal and external temperatures define the actual temperature under which the component being designed is expected to perform. It is possible that the external temperature will be above that specified above, but that the casting will carry either a liquid, air, or gas that is well below the limit. In such a case, the performance heat requirement will be somewhere between the two. A thermal analysis, either mathematical or finite element, is necessary to quantify the numerical heat level.

Structural requirements must be calculated from the expected loading and allowable deflection. Physical properties of the various casting alloys available for die casting must then be compared to the performance calculations. Flexure formulae or finite element programs may be utilized for analysis of the following different loadings:

- *Continuous* loading may induce creep or stress corrosion cracking in some casting alloys.
- *Intermittent* loads describe peak stresses that determine the shape and mass of design in local regions of the part.
- *Cyclic tension and compression forces* or successive deflections introduce fatigue, which is another serious load factor that has to be quantified and compared to the physical property of the casting alloy.
- *Impact* force can cause gross distortion or even fracture if it is not carefully calculated and resisted by the product design.

The common boundary between the part being designed and other components presents a possible design advantage that may reduce both casting and tooling costs and improve reliability. It is therefore important to examine this opportunity, but there are some issues to consider that may affect the performance of the assembly:

- *Dissimilar materials* suggest special design strategy, particularly at the points of attachment, if the thermal coefficients vary enough to cause temperature variance. There is also the potential for galvanic action that will result in corrosion.
- *Galvanic corrosion* is an electrochemical reaction that occurs at the interface of dissimilar metals of an electrolyte. An electrolyte is a liquid that is capable of conducting electricity.
- *Galvanic corrosion deteriorates anodic metals*, especially with a spread on the electromotive scale. The product designer is wise to select metals for interfacing parts that are more compatible. Features that will trap moisture should be avoided to minimize the presence of an electrolyte. A moisture proof barrier is also suggested.

The severity of corrosion is a function of the relative positions of the metals in the electromotive series that follows:

	Electromotive series		
Anodic	Magnesium		
	Beryllium	Aluminum	
	Manganese	Zinc	
	Chromium	Iron	
	Cadmium	Nickel	
	Tin Lead		
Neutral	Hydrogen	Copper	Mercury
	Silver	Palladium	Platinum
Cathodic	Gold		

Fastening die casting components to mating parts is achieved by various methods. Usually, threaded fasteners like bolts, screws, and nuts are employed. Sometimes, it is possible to self-tap or thread a die cast detail, but this depends upon the ductility of the casting alloy. Where more tensile strength or bearing capability than the die cast alloy provides is called for, inserts are either cast or pressed into the casting.

It is sometimes possible to die cast external threads when the parting line is parallel to the longitudinal axis of the thread. In this case, it is customary to truncate the thread so that the trim edge is flat rather than saw-toothed.

Limited use of crimping, staking, and adhesive bonding is possible, but soldering is almost never utilized. In aluminum, it is possible to weld two castings together, but the region of the weld must be free of porosity, a condition that is not too predictable.

An attempt is made in Table 5 to quantify the envelope for the most popular die casting alloys.

The weight limits are relative to the specific gravity of each alloy. Thus, a 100 lb aluminum casting will be approximately 2. Three times the physical size of a zinc part.

Acceptable variation of specific dimensions of the die casting from mean dimensions called for by the design should be determined only by the fit and function requirements of the end product. It must be understood, however, that tighter allowed tolerances call for more expensive tooling, a higher degree of manufacturing difficulty, and higher piece costs.

Table 6 outlines the limits that each casting alloy offers.

Normal commercial die casting production is accomplished at the most economical level and tolerances are referred to as "standard." If greater casting accuracy is necessary, extra precision is required in die construction in addition to better control of the production process. Such tolerances are called "precision."

Factors that influence dimensional deviation from the design mean are mainly thermal in nature. The distance of a particular detail from the shot center must be considered because that is many times the hottest spot in the die. The mass of the detail significantly affects linear tolerances by

Table 5 Limits of Various Casting Alloys

		Casting alloy	
Detail	Aluminum	Zinc	Magnesium
Maximum weight	100 lb	100 lb	65 lb
Minimum wall thickness			
Large castings	0.060 in.	0.030 in.	0.080 in.
Small castings	0.035	0.015	0.035
Minimum draft			
Inside cores	0.035 in./in.	0.017 in./in.	0.025 in./in.
Outside walls	0.017	0.008	0.012
Minimum diameter			
For cored holes	0.094 in.	0.032 in.	0.094 in.

the degree that the temperature of the casting alloy at the end of fill differs from the rest of the shape. The difference in die surface temperatures immediately after ejection between die halves has a definite bearing on dimensional deviation.

Of course, the variance in the die surface temperature from the average over the foot print of the cavity may be critical. A die whose ejection temperature varies more than 20% from the hottest to the coldest points cannot be expected to hold the as cast tolerances described in the standard (Tables 6 and 7). The deviation changes to less than 10% for the precision tolerances.

Basic linear tolerances, in inches, where the dimension is located within a single die component describe the lowest degree of difficulty and are defined in Table 6.

With competition tightening constantly for both price and quality across the whole gamut of original equipment manufacturer to sub-assembly supplier to a single die casting machine, an effective product designer has to use extreme caution in specifying only dimensional tolerances that are absolutely necessary for the end product to function.

Tolerances of dimensions across the parting line are another variation that must be considered and are stated as "plus" tolerances only because the die closed position defines the lower limit of the tolerance because the dimensions cannot get smaller. These tolerances are necessitated by the

Table 6 Commercial Tolerances

Length of dimension	Die casting alloy			
	Zinc	Aluminum	Magnesium	Copper
Basic linear tolerances (standard)				
Basic tolerance (up to 1 in.)	±0.010	±0.010	±0.010	±0.014
Additional tolerance for each additional in. over 1 in.	±0.001	±0.001	±0.001	±0.003
Basic linear tolerances (precision)				
Basic tolerance (up to 1 in.)	±0.002	±0.002	±0.002	±0.007
Additional tolerance for each additional inch over 1 in.	±0.001	±0.001	±0.001	±0.002

injection pressure during cavity fill, which forces the dies to blow. They are a mathematical function—the product of the projected area of each cavity times the injection pressure applied to the liquid casting alloy at the plunger tip for either hot or cold chamber operations.

Both standard and precision tolerances that are to be considered in addition to linear tolerances are given in Table 7.

Special die casting machines are able to hold tighter tolerances on very small or miniature castings. Artificial aging of zinc castings usually contributes to tighter dimensional control, especially where secondary machining is involved, by accelerating the creep (growth) characteristic of this alloy.

Other dimensional tolerances are available in reference texts, that cover moving die components, draft, flatness, cored holes, cut threads, etc. (NADCA, 1994).

Machining stock must be provided for in cases that require more accuracy than can be achieved by the die casting operation alone. The term "near net shape" as opposed to "net shape" is used to describe this condition. To avoid excessive tool wear, a minimum of 0.010 in. is recommended and the

Table 7   Tolerances Across Parting Lines

Projected area of single cavity	Die casting alloy			
	Zinc	Aluminum	Magnesium	Copper
Parting line tolerances (standard)				
Up to 10 in.^{2}	+0.0045	+0.0055	+0.0055	+0.008
11–20 in.^{2}	+0.005	+0.0065	+0.0065	+0.009
21–50 in.^{2}	+0.006	+0.0075	+0.0075	+0.010
51–100in.^{2}	+0.009	+0.012	+0.012	–
101–200 in.^{2}	+0.012	+0.018	+0.018	–
201–300 in.^{2}	+0.018	+0.024	+0.024	–
Parting line tolerances (precision)				
Up to 10 in.^{2}	+0.003	+0.0035	+0.0035	+0.006
11–20 in.^{2}	+0.0035	+0.004	+0.004	+0.007
21–50 in.^{2}	+0.004	+0.005	+0.005	+0.008
51–100 in.^{2}	+0.006	+0.008	+0.008	+0.009
101–200 in.^{2}	+0.008	+0.012	+0.012	+0.010
201–300 in.^{2}	+0.012	+0.016	+0.016	–

sum of this minimum, the casting dimensional tolerance plus the machining tolerance, determines the total necessary machining stock.

Remember that the best mechanical properties of a high pressure die casting are found at or near the surface. Therefore, any more than a minimum machining stock should be avoided so the machining will only minimally penetrate the less dense zone.

Draft allowance is necessary for all die cast surfaces that are oriented perpendicular to the parting plane so that the solidified casting may be ejected from the permanent steel die.

Draft is usually expressed as an angle, which varies with the casting alloy depth of the surface and the type of surface. As the depth of a feature increases, the draft requirement decreases.

Twice as much draft is required for inside surfaces than outside walls because the casting alloy reduces in volume (shrinks) as it solidifies. This shrinkage causes the casting to adhere to the inside die components (usually located in the ejector die) and allows it to pull away from outside cavity surfaces (usually located in the cover die).

It is common practice to specify draft in a general note that is acceptable for the total net shape to be die cast, with inside draft being the deciding factor. However, it is also quite normal to define less draft for special details, but this design decision should not be made arbitrarily. Mathematical formulae have been developed that permit a product designer to quantify draft allowances that will be compatible with the die casting process.

If a shape is to be cast from an aluminum alloy and the dimension above the parting line is 4 in., the draft in inches D equals 4/30 or 0.013 in. The draft angle then becomes 0.013/0.01746 or 0.76°. These calculations quantify drafts that can be processed under commercial economic conditions. Formulae that have been accepted by the industry to calculate draft in inches and degrees are described,

Calculation for draft in inches: $D = L/C$
Calculation for draft in degrees: $A = (D/L)/0.01746$
where D = draft in inches; L = length of feature above or below parting line; C = constant, based upon type of feature and casting alloy; A = draft angle in degrees

Table 8 Values of Constant C

Alloy	Inside wall for dimension	Outside wall for dimension	Total draft of hole for dimension
Zinc	50	100	34
Aluminum	30	60	20
Magnesium	35	70	24
Copper	25	50	17

but these do not define the best possible result. Precision draft tolerance is approximately 85% of that calculated by the above formulae, but the reference text should be followed for exact calculations (NADCA, 1994).

The severe competition between die casting and other processes such as plastic injection molding and constantly diminishing labor costs are the driving forces behind the increasing complexity in product designs. It is not uncommon for a new model to require a die casting that is a combination of four or five separate parts that were acceptable on the model that it replaces.

This leads to more challenges for the die casting process. Wall thicknesses that are not uniform present a varying surface-area-to-volume ratio, and when details get excessively thin or disproportionately massive, casting defects are invited.

Sometimes, just the sheer size of a casting becomes a problem to solve because the distance that the liquid casting alloy must travel between the gate orifice where it enters the die cavity to the farthest extremity is very long (greater than 12 in.). It is this writer's experience that casting alloys developed for high pressure die casting cannot travel much more than this before the percent of solidification exceeds 15%, which announces poor fill, cold shut, and nuclei for gas porosity. Large automotive aluminum transmission cases are an example. In this sense, the die casting process can be considered size sensitive.

Complex core arrangements provide obstructions that introduce metal splashing and excessive turbulence within the die cavity during cavity fill.

Another common conflict involves cored holes around the periphery of a near net shape to be die cast. These cored holes are designed as bolt holes to provide convenient assembly to the mating part. Since they are located very close to the edge of the part, they are adjacent to a desirable gate location. Gates located here direct metal streams that will collide with such cores in the die cavity at super fast speeds exceeding 1200 in. per second. The turbulence thus created will disorganize orderly cavity fill to the detriment of the structural integrity of the casting.

In addition, the collision of super heated metal traveling so fast will cause aluminum alloys to solder onto the cores that are in the way and they will break too soon. Therefore, die casters instinctively avoid gates that are too close to cores. In doing so, casting quality is compromised. Usually the product designer is not even aware of it because the mechanics of assembly are considered so sacred that they are not normally challenged. This scenario could be avoided by designing the casting die at the same time as the near net shape.

It is important to remember, though, that if these conditions are quantified or mathematically defined, they may be analyzed for predictable results. Many times it is possible to overcome such interruptions without affecting the part design by incorporating transfer bridges that act as internal runners.

Many times a die casting is referred to as a geometrical shape, and each shape family calls for its own unique fill strategy.

The flat plate is easy to fill, but then there are not many flat plates to be die cast. This is a very simple shape to die cast unless a tight flatness tolerance is specified.

The box shape must be dealt with carefully because the liquid metal streams will travel in a straight path if there are no obstructions. However, the walls nearest the gate will see the best quality at the expense of the far side. This depends, of course, upon the distance that the casting alloy must travel during cavity fill, as shown in Fig. 13.

The hat shape is difficult to fill because the metal stream tends to run around and fill the vertical side walls before the

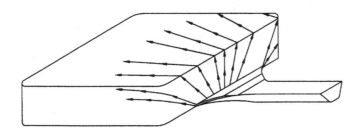

TYPICAL FILL PATTERN FOR BOX SHAPE

Figure 13

top is back filled with no place to vent out air that then becomes entrapped. The flow vectors in Fig. 14 illustrate this.

    The cylinder presents another set of challenges since the inside diameter is usually formed with a long minimum draft core. This shape can be cast in the lay down position or standing up in the die.

    The boomerang shape sometimes presents an uneven projected area and it is difficult for the die casting machine to hold the die halves together. The real significance of this shape is that it calls for gate locations on the inside edge where the metal streams may diverge to fill more of the cavity by fanning out. If the outside is gated, the metal streams converge upon a central location for very solid fill at the expense of the rest of the shape. The flow vectors are shown in Figs. 15 and 16.

    A particular advantage of the die casting process is that intricate coring of holes, slots, or any depressed detail is easily accomplished. There is an additional tooling cost which can be

SINCE TOP IS BACK FILLED AND CAN NOT BE
VENTED, GAS AND AIR ARE ENTRAPED AT TOP

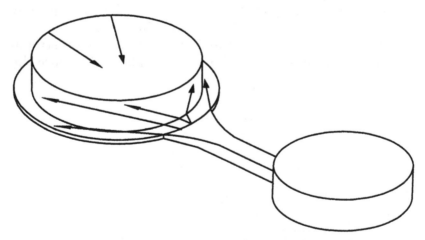

TYPICAL FILL PATTERN FOR A HAT SHAPE DESCRIBES INABILITY
TO FORCE METAL STREAMS UP AND OVER TOP SO FLOW VECTORS
FOLLOW THE PATH OF LEAST RESISTANCE WHICH IS AROUND THE
LIP AND SIDES

Figure 14

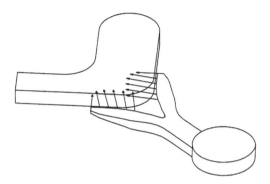

Figure 15

amortized over a large quantity of parts. Thus it is possible to cast a near net shape which many times is the economic jus-tification for die casting a part.

There are some rules that relate to the slenderness ratio (length/diameter) of steel cores in the die that, if followed, can minimize mechanical production problems. This is also a function of the behavior of the casting alloy as illustrated in Table 9.

Interlocking cores may be used in extreme situations and can be practical but it must be understood that more heroic die maintenance is called for as well as extra careful production

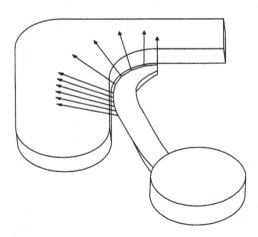

Figure 16

Table 9　Slenderness Ratios for Cores to Minutes Replacement

	1/8	5/32	3/16	1/4	3/8	1/2	5/8	3/4	1
	Diameter of cored hole in inches								
	Maximum depth in inches								
Alloy									
Zinc	3/8	9/16	3/4	1	1 1/2	2	1 1/8	4 1/2	6
Aluminum	5/16	1/2	5/8	1	1 1/2	2	1 1/8	4 1/2	6
Magnesium	5/16	1/2	5/8	1	1 1/2	2	1 1/8	4 1/2	6
Brass				1/2	1	1 1/4	2	3 1/2	5

techniques. Remember that one of the interlocking cores must pull out before the other can move, usually by the dies opening. Otherwise, the weakest core will break to generate downtime to replace the broken die component. Eventually such a condition will increase costs when all parties recognize it.

A typical interlocking core arrangement is described here. In this example, the movable core is mounted in the ejector die and must be pulled out along its center line *before the die can be opened.* Such cores eliminate the cost of secondary machining but tend to slow down the casting cycle. Thin flash, especially where aluminum alloys are cast, builds up between the movable core and the way that it must move in, since liquid metal is injected adjacent to the core and the way in. Fig. 17 details this condition

Submarine cores are required by holes that are parallel to the parting plane but are offset. Similar die maintenance to the interlocking core is required in this case, since the core must be pulled *before the dies can be opened* and thin flash tends to build up around the core. Many times, it is impossible to remove this flash without forcing the core out of its mount because it moves in an opening that lies below (submerged) the parting plane.

An insert is a component, usually manufactured from a material that displays different properties than the casting alloy. It is not produced in the same die casting cycle, but sometimes may be another die casting. Design requirements,

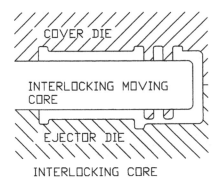

INTERLOCKING CORE

Figure 17

in addition to performance, are that the insert be cast firmly in place with no subsequent movement. Usually the casting alloy shrinks onto the insert to achieve this objective.

There are several reasons for incorporating inserts into a die casting. Inserts are used as fasteners to mating parts to eliminate separate bolts, pins, screws, or welds. Inserts impart properties that are not inherent to the casting alloy like hardness, mechanical strength, abrasion and corrosion resistance, resiliency (i.e., spring), and magnetism, to name a few advantages.

Power can be transferred through a hardened steel insert cast in place such as a thread or gear.

Ribs are structural features that are used by the astute product designer to increase tensile strength and stiffness, control warpage, and to act as sinks to dissipate excess heat. Ribs also act as feeders to facilitate the flow of liquid metal streams into remote zones of the casting during cavity fill.

Figure 18 illustrates how ribbing can be useful in controlling warpage caused by uneven thermal conditions that occur during the casting process. This is typical of large flat areas. Sometimes warpage can be predicted prior to making the first shot from a new die. There are distortion computer models available to make complicated calculations. However, usually warpage is corrected after actual casting experience with a particular net shape.

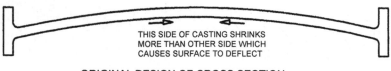

ORIGINAL DESIGN OF CROSS SECTION

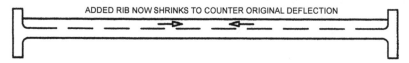

RIB ADDED TO CROSS SECTION DESIGN

Figure 18

Heavy walls may be lightened considerably and at the same time increased in strength by the addition of ribs. Thin ribs are sometimes composed totally of the dense metal at the surface and are free of porosity.

Die blow is a condition that cannot be avoided because the high injection pressure that generates it is essential to the casting process. Under normal pressure, which translates to approximately 5000 psi over the projected area of the cast shot, the die halves are forced apart by a distance of 0.01 in., which is considered reasonable. Of course, under unusual circumstances such as insufficient locking force, the separation between the die faces is more.

This movement, described in Fig. 19, also occurs in the case of core pulls normal to the die parting plane; dimensional changes of similar amounts must be expected.

Parting line flash develops as a result of die blow and must be removed from the as cast part to comply with the product design. Hydraulic or mechanical trim dies are employed to perform this operation. Fig. 20 describes a typical part trim scenario.

Though it is referred to as trimming and is sometimes accomplished by shear action, this operation is really punching where the product is supported by a steel die, and the punch travels through the flash plane. Sometimes the action

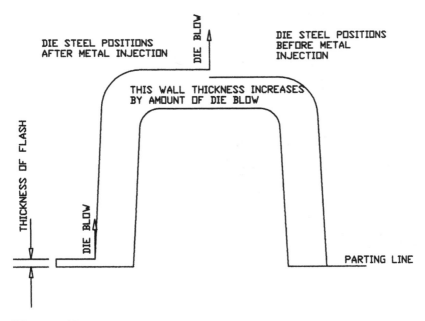

DIE STEEL POSITIONS
AFTER METAL INJECTION

DIE BLOW

DIE STEEL POSITIONS
BEFORE METAL
INJECTION

THIS WALL THICKNESS INCREASES
BY AMOUNT OF DIE BLOW

THICKNESS OF FLASH

DIE BLOW

PARTING LINE

Figure 19

is the reverse in that the part is punched, as in the case of
drop through dies. It is important that the product design
allows enough space for sufficient follow through of the trim
cutters.

The contour to be trimmed can impact both tooling and
production costs, so the more simple it can be the better. A
typical simplification of a complicated parting line configuration is illustrated in Fig. 21.

Lettering or any form of artwork can be cast on the surface of a die casting. The lettering may be raised or depressed.
Raised lettering is the most economical since the figure can be
cut directly into the die steel while depressed letters must be
left standing on the cavity surface while the material around
it must be removed.

Sometimes, because of fit with a mating component,
raised letters are not acceptable. In this case, the lettering
is raised but within a depressed pad to satisfy the fit requirement and also minimize tooling cost.

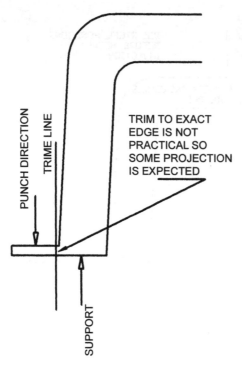

Figure 20

A minimum width of the top surface of each letter of 0.01 in. is recommended plus 10° draft per side all around. The height of each character must be equal to or less than the top width.

Several design formats are regularly used to convey graphical information between the die caster and the product designer. *Hard copy paper* drawings are still popular and can be universally used by all die casting engineers and tool makers.

Electronic files that are computer generated by several computer-aided design programs are also used extensively in the die casting industry. Some of the programs used are Pro Engineer®, Auto Cad®, Cad Key®, and many others too numerous to mention here. Such files are exported in the IGES (International Graphics Exchange System) format that can be universally read and imported into any CAD system.

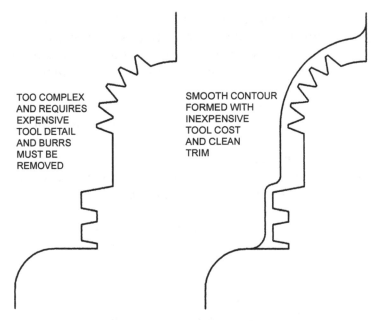

Figure 21

These files can be either two- or three-dimensional. Most die casting firms have CAD skills and knowledge and tool makers are even more sophisticated in this regard.

The economic dimensions of the die cast component are critical to the design procedure. It helps to know that 65% of the cost is in the casting alloy, 20% is in manufacturing labor and burden, 5% in melting energy, and 10% in selling, and general administrative expenses.

The largest cost element, metal, is sold by the pound, usually based upon the applicable metal market price, but consumed by volume. Therefore, it is worth considerable investment in design time and talent to thoroughly analyze the volume of metal required by the design. Care must be taken, however, to monitor the ratio of volume-to-surface-area after each design move so that reasonable quality can be expected in production quantities.

Tooling cost is amortized in one way or another over the quantity of parts that are expected to be produced during the

design life of the part. Since tooling for die casting is manufactured from special steels and intricate shapes are involved, its
cost is usually a major economic factor. Therefore, the anticipated volume of consumption has a serious impact upon the
economic decisions.

The degree of difficulty to produce the net shape that the
product designer creates is a critical economic ingredient.
There are so many tooling cost-cutting measures that it is
not practical to cover them here. It should be noted, however,
that almost every one introduces a disadvantage in production. Some are acceptable if the anticipated volume does not
justify more robust tooling.

This subject is covered in some detail in chapter 12 on
mechanical die design where the cost to performance is
compared.

Three-dimensional models that eliminate the need for
paper designs are also frequently used and are an excellent
method for communicating with the die casting industry.

Product design is the main medium of communication
between the final product function and appearance and the
manufacturing process. The typical designer is not an
experienced die caster and, while an attempt is made here
to explain the most essential characteristics, it is not possible
to discuss all of the diverse skills and knowledge that the professional die caster has acquired.

Therefore, ideally the die caster that will actually produce the product is selected early so the project engineer
may enter the product design procedure almost as a technical
partner. It is at this stage when many positive suggestions
might be incorporated into the product design. This is not
usually the case, however, because the bidding process to find
the lowest piece price and tool cost does not take place until
the product has been designed. However, what does frequently occur is that certain die casters specialize in particular products and become more intimate with the technical
and business culture of their customers. Some examples are
automotive power train parts like valve bodies and transmission cases, computer components, hardware finish cosmetic
requirements, etc.

Short of simultaneous design of both the product and the die casting die, full and complete information should be provided to the die caster. Some suggestions are offered here:

Identify all locations where secondary machining is expected, with finished dimension and tolerance, so that necessary machining stock may be provided.

Indicate clearly plus / minus dimensional tolerances required in the net shape to be die cast.

Specify direction and maximum permissible draft on all wall sections that will satisfy fit and function of the finished product.

Define any special requirements that may not be usually considered in commercial standards, some of which follow:

- Leakage resistance and the leaking medium of air, gas, water, oil, etc.

- Pressure (high or low) other than atmospheric.

- Locations that must be free of porosity.

- Locations that must be free of surface marks from ejector pins, parting line flash, etc.

- Isolated flatness requirement and tolerance allowed.

- Cosmetic surfaces and ultimate finish to be applied.

- Location and nature of strength requirements and their nature: bending, torsion, twisting, tension, hardness, etc.

- Location of surfaces that are exposed to corrosion like sea water, humidity, contact with fumes, chemicals, foods, etc.

The effective product design phase precedes the tooling and production operations by enough time to somewhat distance the product designer from manufacturing, so a general summary of this chapter is listed below:

- Casting alloy selection affects the function of the product and the liquid metal flow during cavity fill.

- The parting line is established by the design of the foot print of the part. Cost and quality are a direct function of the configuration.
- Provide minimum wall thickness consistent with adequate strength and stiffness requirements.
- Keep uniform wall thicknesses.
- Maintain an acceptable ratio of volume-to-surface-area to address cavity fill requirement.
- Design shallow ribbing to minimize distortion on large flat surfaces.
- Avoid cored holes parallel to the parting line that are not on the parting line.
- Avoid undercuts.
- Allow adequate draft.
- Be cautious about passages that require interlocking cores.
- Balance the cost of casting difficult details against producing them by secondary operations.
- Core out metal savers wherever possible to minimize massive details.
- Provide adequate space between cored holes to accept the most robust die design.
- Conform to slenderness ratio specifications for long cored holes.
- Try to keep cored holes either perpendicular or parallel to the parting plane.
- Utilize ribbing to reduce the incidence of stresses that are expected during solidification.
- Provide adequate draft.
- Specify uniform radii and fillets to break sharp corners which reduce die life and increase tool cost.
- Be careful with deep projections that trap air and may obstruct flow of the casting alloy.
- Consider the use of an insert to be cast into the part where properties different than the casting alloy are necessary.
- Specify raised lettering for tool economy.

- Provide more tolerance on dimensions across the parting line.
- Estimate the cost of producing a complex shape as a group of separate parts and also as one part so that total manufacturing costs may be compared.

3

The Die Casting Machine

The history of the die casting machine takes us back to the time of the great American gold rush of 1849, following the discovery of the treasured yellow ore at Sutter's Mill, California. Thousands of pioneers called "Forty-Niners" went west to dig, pan, scrape, and scramble in search of the elusive precious metal.

A continent and an ocean away, another pioneer, named Sturgiss, began the rush for a more common metal ... yes, this started the "lead rush." In the same year, 1849, Sturgiss patented the first die casting machine, except it was called the "lead kettle," designed to cast printer's type.

For centuries alchemists had tried in vain to change lead into gold through mysterious and esoteric, and even occult procedures. The proper formula always evaded them. But with the introduction of the Sturgiss machine, illustrated in schematic in Fig. 1, the value of lead was greatly increased without any change in its atomic structure. This machine made casting of liquid metals quick, efficient, economical, and repeatable. *The "golden age" of die casting had begun!*

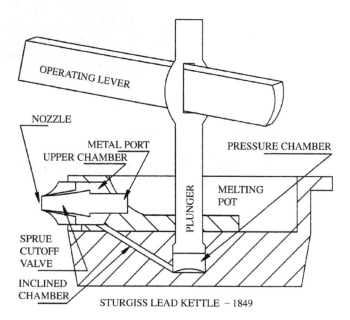

Figure 1

Though you may not realize it, the Sturgiss die casting machine is identical in its function to modern machines. If you look closely, it is the forerunner of the hot-chamber process. However, there is an enormous difference between the "lead kettle of 1849" and one of today's automated wonders.

The early die casters had no respect for the limitations of their machines, much like their modern counterparts of today. If a certain task could not be performed, they modified the machine until it could. This was not aimless tinkering, but continuous and steady improvement to allow the casting of larger and more complex shapes.

Some changes were minor, as in the improvement described in Fig. 2, but still worthy of being patented. Metal was poured into the plunger cylinder through a port (not shown). The liquid metal was forced out through the nozzle and into the die by a sharp blow on the wooden knob. Then spring pressure returned the plunger to its upward position.

By 1877, C. and B.H. Dusenbury had invented a machine similar to the Sturgiss machine, but different

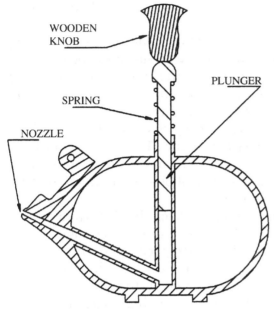

WOODEN KNOB

PLUNGER

SPRING

NOZZLE

SIMILAR TO MODERN GOOSENECK

Figure 2

enough to qualify for a patent. This machine was designed with a hollow plunger containing a valve which allowed the liquid metal to flow from the upper to the lower chamber. Rather than a sprue cutoff valve, the Dusenburys utilized a movable die that was held against the nozzle until the cavity was filled, and then moved away to break the shot off from the sprue.

The casting of printing type was almost the only application for die casting for a couple of decades after the introduction of the Sturgiss machine, but by 1870, a small machine that was capable of casting other small shapes was in operation. Die casting progressed to the production of parts for cash registers and phonographs.

The casting alloys used were of lead and tin because of their low melting points and fluidity. However, the applications were limited since both metals are soft and neither is strong. Furthermore, tin was and still is expensive, but the consumption of these alloys continued until about 1920.

During this time, the automobile came on the scene nd by 1904, the H.H. Franklin Company was die asting bearings for connecting rods. In that year, the automobile industry replaced the printing industry as the primary user of die castings, position it has held ever since.

The development of die casting machines has paralleled the consumption of die cast parts since 1904. Since that time, die casting machines have evolved from a primarily manual operation, through various stages of power operation, to some very sophisticated automation.

Aluminum was also being cast for the first time around the early 1900s, and this lead to significant changes in die casting machine design. The melting point of aluminum is high compared to lead, tin, or even zinc. In the die casting operation, molten aluminum corroded the iron and steel parts of the machines that caused a high mortality rate. The castings produced were also contaminated with iron inclusions. It was concluded that the machine plunger could not remain in direct contact with this alloy during injection.

This stimulated the invention of the gooseneck in 1907 by Van Wagner as described in Fig. 3. Van Wagner's design used air pressure rather than a plunger to inject the liquid aluminum into the die. The gooseneck was fixed and was filled by hand ladling through its nozzle outlet. Once the gooseneck was filled, the die was rotated 90° and locked in place over the nozzle. The air pressure was then applied through the air line, forcing the liquid aluminum up the gooseneck and into the die cavity.

Later gooseneck machines were usually arranged like the schematic illustrated in Fig. 4, where the gooseneck is immersed into the molten metal and if filled, it raised up to the die before it is pressurized. The die is then filled in the horizontal direction.

Gooseneck machines operate at rather low pressures and have large iron surfaces in contact with the molten aluminum. This process has therefore given way to cold-

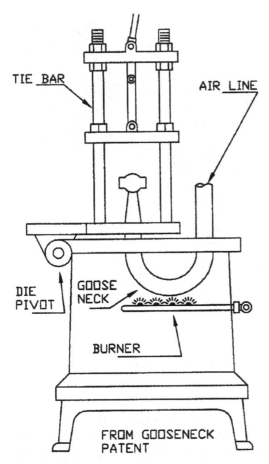

Figure 3

chamber machines where liquid aluminum is ladled into the cold chamber (shot sleeve). This did not occur until the 1930s and their use has increased steadily until today.

Entering the modern era, two types of die casting machines emerged — the cold chambered and the hot chambered. The main difference is in the way the molten metal is delivered to the metal feed system of the die.

Hot-chambered machines are primarily used with alloys of low melting points (less than 800°F) like zinc, lead, tin, etc. However, the hot-chambered machine is also widely used to

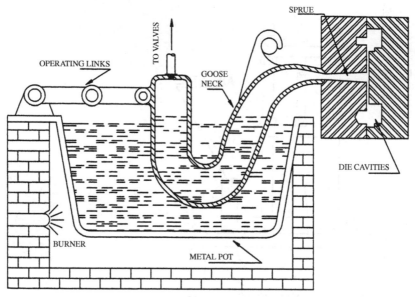

GOOSENECK IMMERSED IN MOLTEN METAL

Figure 4

cast the magnesium alloys at 1200°F since is does not have an affinity to dissolve iron. Cold-chambered machines are used with higher melting point metals like aluminum and brass.

The hot-chambered type is much more efficient in that the superheated metal is forced hydraulically into the die where the cold-chambered process requires ladling of the casting alloy from the holding furnace into the pour hole of the cold chamber. Of course, automatic ladling mechanisms have been in common use since the 1960s, but the procedure still adds time to each casting cycle. Also, it is not difficult to observe hand ladling over a broad portion of the industry.

Even though this additional handling of the alloy expends time, the more technical objection to ladling is the thermal compromise that occurs when the column of liquid metal is poured in air. Heat is lost so the melt of metal must be held usually at approximately 50°F above the desired injection temperature. In addition to heat loss, oxides are also introduced into the casting alloy just before it enters the die impression to form the casting.

The reaction of the high temperature casting alloys upon the materials of the shot system mentioned above is the reason for development of the cold-chambered process. Yes, inert materials like ceramics have been experimented with, but not successfully.

The early die casters made their own machines. It was not until the Soss Manufacturing Company placed its machines on the market in the early 1900s that die casting machines became commercially available. The Soss machine had a patented sleeve-mounting system that advanced the art of die casting, but it is best remembered for being the catalyst that helped the fledging industry to take wing. Undoubtedly, the availability of workable die casting machines, at a time when die casters jealously guarded their developments, helped to attract more custom and captive shops to the die casting process.

Perhaps the first automatic die casting machine marketed in the United States was introduced by Madison-Kipp Corporation in 1928. This hot-chambered zinc machine had a 10 by 14 in. spacing between guide bars. In 1930, the company offered more advanced units that included one model with a 12 by 16 in. guide bar spacing.

Cold-chambered attachments for the machines were available by 1932, so die casters could die cast aluminum and brass. In late 1931, Kipp engineers were working with Dow personnel on the development of a sulfur dioxide dispensing device and metal pot cover that could be used in die casting *magnesium* ... probably the first magnesium die castings produced in North America!

Another, less widely used configuration of die casting machinery is the vertical, as opposed to horizontal machine. It was developed in the 1960's to reduce the potential for air entrapment in the metal feed system. Figure 5 shows a schematic of the vertical machine. The short plunger is vertical, rather than horizontal and travels up to the dies.

Figure 6 compares the tendency to encapsulate the air in the shot sleeve into the liquid casting alloy via the usual horizontal cold chamber method to the vertical strategy, in which the air is injected into the cavity ahead of the liquid metal stream. This concept is covered in detail in chapter 7 where the metal feed system is discussed.

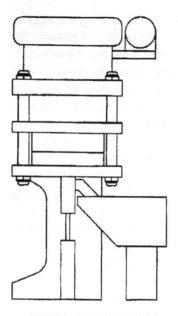

VERTICAL DIE CASTING MACHINE

Figure 5

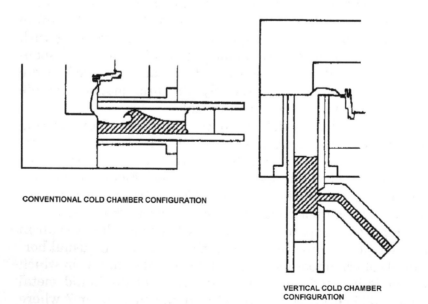

CONVENTIONAL COLD CHAMBER CONFIGURATION

VERTICAL COLD CHAMBER
CONFIGURATION

Figure 6

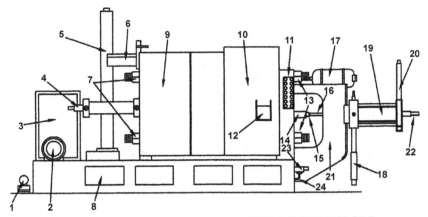

VIEW OF MODERN COLD CHAMBER MACHINE FROM THE OPERATOR SIDE

Figure 7 Key components are numbered and identified in Table 1.

Table 1 Key components

Item No.	Description of component
1	Heat exchanger (sometimes referred to as after cooler)
2	Main motor
3	Electrical cabinet
4	Die lock cylinder
5	Die lock accumulator (piston type illustrated)
6	Safety latch mechanism
7	Tie bar nuts
8	Access cover to reservoir (typical)
9	Linkage guard
10	Safety gate
11	Operator's control station
12	Observation window
13	Tie bar nuts
14	Cold chamber (sometimes referred to as shot sleeve)
15	Plunger tip
16	Shot arm
17	Shot nitrogen accumulator
18	Jacking mechanism to change shot end positions (center or below)
19	Shot cylinder
20	Shot accumulator (piston type)
21	"C" frame that supports shot end
22	Shot stroke adjustment
23	Shot speed control
24	Hydraulic return line

In the vertical system, venting of the air introduced into the cavity is more critical. Vacuum is usually used to reduce the volume of air in the cavity during cavity fill.

The logic of the vertical injection strategy is so obvious that one would think it would be in almost universal use, given the quality problems with gas porosity. However, only a very few die casting firms have embraced this concept with good success. Since vertical cold chamber die casting is not widely used, it is mentioned here to recognize its technical, if not commercial potential.

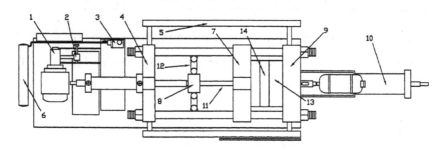

TOP VIEW

MODERN COLD CHAMBER DIE CASTING MACHINE

Figure 8

Table 2   Key Components

Item No.	Description of component
1	Pumps
2	Suction valve
3	Circulating pump and filter
4	Adjustable platen
5	Helper side safety guard
6	Heat exchanger
7	Moving (ejector) platen
8	Cross head
9	Stationary (cover) platen
10	Shot cylinder
11	Cross head guide rods
12	Mechanical locking linkage
13	Cover die
14	Ejector die

The modern die casting machine is schematically illustrated in Fig. 7 in the cold-chambered configuration.

Most die casting machines have the same components. Platens, tie bars, ejection systems, and accumulators, to name a few. Most are powered by electric motors, though a few are still air driven. Another schematic view follows (Fig. 8) that describes other important machine components (Table 2).

The major components will be discussed here in detail so that their functions are made clear.

The basic structural component, the machine base which supports both stationary and moving parts is shown in Fig. 9. Great care must be taken when setting the base because it must be absolutely level to keep the machine from twisting or moving (walking). Most installers use laser transits to level the base within 0.003 in. (0.075 mm). The base has to be firmly fastened to the concrete platform to preclude any detectable movement of the machine base during operation.

The base serves several functions. It is the frame upon which the whole machine rests. The moving (traveling, ejector) platen and the rear (adjustable) platen are sat upon, but not fastened to the base platform. The stationary (cover, front) platen is fastened to the base. The base can also serve as the hydraulic reservoir as a portion of the clamp end of

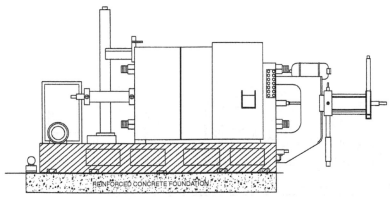

MACHINE BASE IS HATCHED

Figure 9

the base is enclosed to form a tank to contain the hydraulic fluid. The base has to have enough strength to not only withstand the weight of the platens, but should not twist or flex under the high pressures that occur during the cycling of the machine. Twisting or flexing can damage the machine or the die.

The heavy steel fabrication:

• Is a platform for other components;

• Must be strong to avoid bending;

• Must be rigid to avoid twisting.

The platens are also structural components in the form of three large plates of machined steel that support the machine loads and the dies. The stationary platen is attached directly to the base while the ejector platen and the adjustable platen have freedom of movement (slide front to rear, not side to side) upon the base. Figure 10 high lights the platens.

The stationary platen is located at the "shot" or "injection" end of the machine. This platen holds the cover portion of the die within the "die height" space. The shot system is mounted on the other side of this platen.

The movable (traveling, ejector) platen is located in the middle of the machine. The ejector portion of the die is mounted within the "die height" space. The die ejection

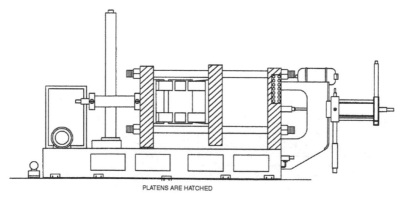

PLATENS ARE HATCHED

Figure 10

system is mounted on the other side, along with the toggle linkage that performs as the mechanical clamp. This platen is usually supported by an adjustable shoe to ensure alignment.

The rear (adjustable) platen is mounted at the rear or clamp end of the machine. The toggle linkage system is mounted to the inside surface of this platen, and the hydraulic closing cylinder and the die height adjustment system are located on the outside surface. This platen is also supported by a shoe that allows back and forth movement on the base.

The movable and the adjustable platens move with every cycle or "shot." The moving or ejector platen slides back and forth to open and close the die. The adjustable platen slides just a small distance as the tie bars stretch while the dies are closed. The adjustable and ejector platens will move during "die height" adjustment. This adjustment raises or lowers the force the die halves generate upon closing.

Perhaps the most important maintenance required on the cover and ejector platens is to keep them clean to ensure that they are parallel and that a good transfer of heat can be expected between the dies and the platens.

T-slots or tapped holes are incorporated into the die mounting platen surfaces so that the dies may be clamped into operating position. Cleaning during every die change is also important to keep them from being damaged. Figure 11 describes a T-slot arrangement in the cover or stationary platen and the relationship to the cold chamber that is set from the "die height" space between the platens.

Safety concerns regarding platens are for burns and crushing of human extremities. Both platens may become hot enough to burn during operation, especially the stationary platen on a hot-chamber machine. Also, by their very nature, the platens may become a snag or strike hazard. Be sure that all safety barriers are in place and safety locks are working at all times when the machine is under power.

Tie bars orient and position the platens. Most machines have four, but some have three, and one machine manufacturer, Lester Machine, replaced the tie bars with a solid frame. Although no longer manufactured, many of these machines are in operation.

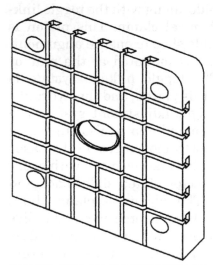

TYPICAL "T" SLOT ARRANGEMENT ON
STATIONARY PLATEN

Figure 11

Tie bars are highlighted in Fig. 12 to conceptually show their relationship to the rest of the machine. The moving platen slides along the tie bars during the opening and closing cycles. The size and strength determine the locking capacity of the machine. The size is also specified by this locking force. These tie bars are utilized in machines that produce a wide

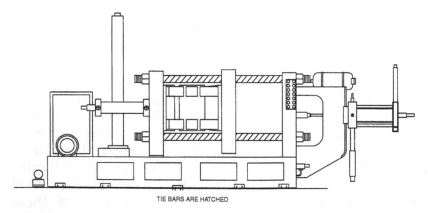

TIE BARS ARE HATCHED

Figure 12

range of castings from a few ounces to more than 80 pounds. Of course, the size of the tie bars is proportionate to the locking force.

The locking capacity for each tie bar is determined by the strain or stretch that it experiences during cavity fill when the full operating pressure is applied upon the liquid metal. A chart is included in Fig. 13 that relates tie bar strain to clamping force for an individual tie bar.

An important datum for the die designer is the center of inertia of the projected area of the total shot including cavity, runner, and overflows. This point should be located as close as possible to the center of the tie bar pattern so that the shot pressure will be evenly distributed. Then, the clamping force required from each tie bar will be equal, or close to it.

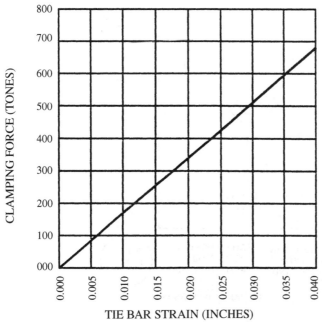

CLAMPING FORCE IS FUNCTION OF TIE BAR STRAIN

Figure 13

Sometimes it becomes necessary to deviate from the conventional wisdom of horizontal and vertical symmetry to properly balance the die.

In Fig. 14, a poorly laid out die exceeds the locking capability of the machine because one of the tie bars must hold too much of the force, while the other three are only required to supply a fraction of their design strength. In this case, the die must be operated in a larger, more expensive machine or risk breakage of the tie bar if the smaller machine is used. The die layout can be designed to accommodate the shape of the shot, but the first decision is made more often than not in actual practice. Note that, even though the die requires only 515 tons of locking force, an 800 ton machine is too small.

When one tie bar is taxed by an unbalanced condition, it usually breaks in front of the threads at the stationary platen,

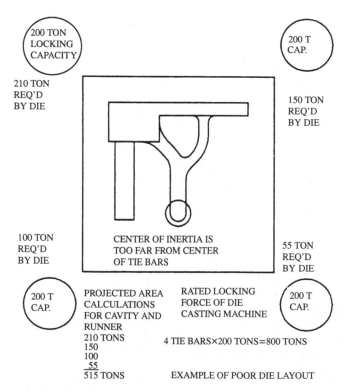

Figure 14

or the threads are stripped. Figure 15 describes one solution to the problem. It is unique because the straight edges of the cavity are rotated at an angle. Although CAD technology and CNC machining will offset additional cost, such a design could increase the cost of constructing the die, but this additional cost is small compared to the production costs that will be incurred later. The problem is that each decision is not normally made at the same time so a clear cost comparison becomes blurred.

The toggle linkage system opens and closes the die halves since it is connected to the adjusted and movable platens. This mechanism may look different on various manufactured machines but still serves the same function. All are designed as a levering mechanism to gain a mechanical advantage. This reduces the requirement and size of the die closing cylinder to still be able to close with very high force.

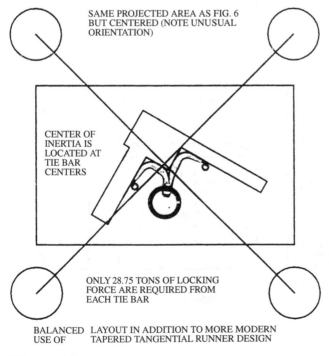

SAME PROJECTED AREA AS FIG. 6 BUT CENTERED (NOTE UNUSUAL ORIENTATION)

CENTER OF INERTIA IS LOCATED AT TIE BAR CENTERS

ONLY 28.75 TONS OF LOCKING FORCE ARE REQUIRED FROM EACH TIE BAR

BALANCED USE OF LAYOUT IN ADDITION TO MORE MODERN TAPERED TANGENTIAL RUNNER DESIGN

Figure 15

A schematic is offered in Fig. 16 that offers an opportunity to study the toggle linkage.

Die height adjustment is accomplished two ways. The simplest method is tightening or loosening the tie bar nuts

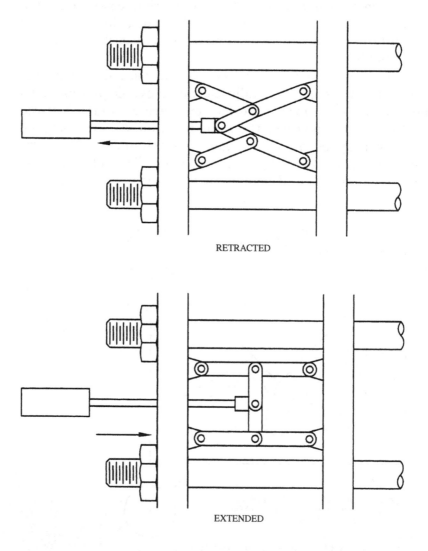

RETRACTED

EXTENDED

TOGGLE LINKAGE SYSTEM

Figure 16

on the rear platen. Remember, tightening the nut stretches the tie bar, which increases the locking force that can be applied by that bar to hold the die halves closed. When this adjustment is done by hand, the strain (stretch) on each tie bar should be measured to ensure equal tension on the locking system. Even though everyone knows that this should be done, the measuring part is too often skipped.

The danger here is that, as the toggles and locking system wear, the tendency is to gradually force the platens out of line so that they are not closely parallel to each other. Thus, this is a very delicate part of each die setup.

Motorized die height adjustment is automatic if no tinkering is involved and usually includes strain gages that are built into the rear end of each tie bar. If this system is properly maintained and the platens stay square, improved internal and dimensional quality will result. A typical motorized system is described here.

There are some serious problems that occur from poorly adjusted tension on the tie bars that occur gradually over several thousand cycles. Because they take place gradually, these problems generate quality problems that are sometimes difficult to diagnose because of their insidious nature. First, the platens of the machine are forced out of square so that they do not close the die halves equally. Of course, the next thing that occurs is the die halves are forced out of square and then they will only run properly in the distorted machine. This is where the human equation comes in — human nature in the person of set up foremen, set up crews, machine operators, or plant management influence the scheduler to only match the die with the single machine. This then creates down-time at other, better machines and eventually adversely affects the bottom line.

The ultimate fix comes when the machine wears to the point that it must be rebuilt, which includes remachining of the out-of-square platens so that they again close properly, even though they are now thinner because of the stock that has been removed. The final blow comes when the distorted die will no longer run effectively after the expensive rebuilding. This chain reaction is exasperating to management and contributes to the reputation that the die casting process is unpredictable.

Obviously, the best strategy is to prevent this alignment disaster from occurring in the first place. The automatic or motorized die height feature is an accessory available from all machine manufacturers at added cost. It should be easy for the reader to see that the added cost here can prevent expensive fixes as the machine wears.

This feature is not fool proof and is subject to wear of the moving parts and normal abuse at the die casting plant. It is important to keep it maintained and not to merely disconnect it since it is not essential to the process.

A common motorized die height device is described in Fig. 17 and Table 3.

Power is generated by electric motors and valves that convert hydraulic fluid into energy. Some machines have only one motor; larger machines have several with varying horsepower depending on the task they must power. The electric motors operate at a high voltage, usually in the 440/480 range. Therefore, the area around the control panel and the motor should be kept as clean and dry as possible to avoid an electric shock hazard.

All machines have at least two *hydraulic pumps*, a high pressure/low volume and a low pressure/high volume configuration. The coupling between the pump and the motor has to be guarded, with frequent inspections since this is an especially vulnerable area.

Solenoid valves are used to control the volume and direction of the flow of hydraulic fluid. A solenoid is an eletromagnet that shifts a metal core. Solenoid valves can be very small (less than a couple of pounds), or they can be very large, weighing close to 100 pounds. Usually, they are very sturdy but should not be used as steps or for supports for tools, etc. The solenoid valve determines the direction of movement and the flow control valve determines the velocity of that movement.

Selection is accomplished by the position of the spool in an operating valve assembly, which either provides an orifice for hydraulic fluid or seals it off. A simple cross-section

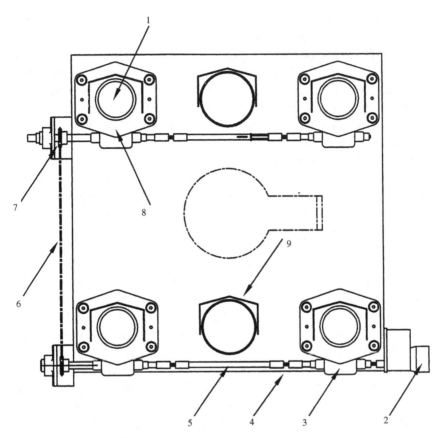

PLAN VIEW OF REAR PLATEN

Figure 17

Table 3   Key Components

Item No.	Description of Component
1	Tie bar position
2	Hydraulic motor
3	Worm gear assembly
4	Universal joint
5	Drive shaft
6	Drive chain
7	Sprocket
8	Die height nut assembly
9	Protective cover

through a solenoid is illustrated in Fig. 18 to describe the opening of one port and the closing of two others.

With the spool in the position shown, ports A and C are sealed off, but port B is open. Therefore, flow will be directed to that system. Thus, the task of directing the movement of flow of the hydraulic fluid is satisfied. All of the parts of the solenoid are precisely machined to close tolerances and the movement is akin to the works of a Swiss watch. The electromagnets are sealed off from the hydraulic fluid, so are immune from any minute solids contained in the fluid, but the spool and channels directly come into contact with it.

This brings to mind the absolute need for cleanliness in the hydraulic system including, but not limited to, the reservoir tank, pipes, all connections, etc. Filters in the range of 8 μm are designed into the system at strategic locations, but somehow dirt manages to get through. Cleanliness of the hydraulic fluid from the central source supply (usually barrels, but sometimes a large container) is essential in a die casting operation. This is where the gremlins come from on Monday mornings, also contributing to the myth that the high pressure die casting process is not predictable.

Contamination of the fluid on Monday mornings is referenced because bacteria grow in the fluid that was no problem when it was in suspension in active production. However, when the machine is idle over the weekend or down for some other reason, the bacteria settle causing some microvalving to malfunction. Spermacides are available to minimize this

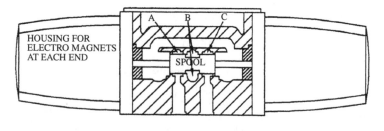

SOLENOID VALVE

Figure 18

effect, but experience with the phenomenon in the maintenance department is essential.

The pilot operated check valve is used to provide an automatic accumulator drain when the pump is turned off. The projected area of A in Fig. 19 is ten times larger than the area of B when the check is seated. When the pump is running, pressure will enter both A and B. Due to the pressure imbalance, the check will remain seated, allowing no flow through the valve, and the accumulator will charge.

In the next scenario, the pump is turned off, and the pressure in area A will drop to zero. Then the pressure at area B forces the check off its seat as given in Fig. 20. The accumulator will then discharge when the hydraulic fluid bypasses the opened check and enters area C. This circuit carries it back to tank.

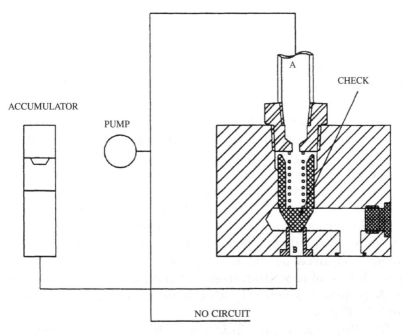

PILOT OPERATED CHECK VALVE [ACCUMULATOR CHARGING]

Figure 19

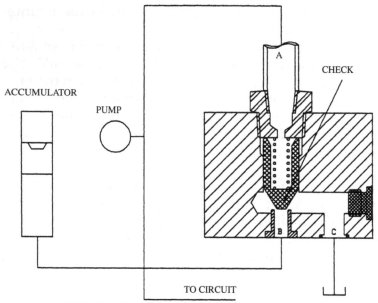

PILOT OPERATED CHECK VALUE [ACCUMULATOR DISCHARGING]

Figure 20

The majority of die casting machines will also have manual or hand valves. These are usually opened all the way in the case of modern machines, which are equipped with computerized process controllers. The settings can be very precise. On older machines without process controllers, it is crucial that the valve settings be accurate, therefore the dry shot plunger velocity related to the number of turns of the valve should be frequently checked. The setting will change under constant use. These machines represent the "old, seat-of-the pants" era. The number of turns on the shot valve philosophy just will not produce the quality or productivity numbers required by die casting users. Valve wear causes too much lack of repetition from production run to production run. A typical flow control valve is described as a schematic sketch in Fig. 21.

The flow of hydraulic fluid through the valve is increased when the needle is lifted off its seat, as the adjusting knob is turned counter clockwise. Conversely, flow is reduced as the

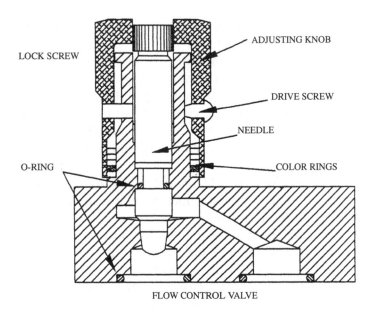

LOCK SCREW

ADJUSTING KNOB

DRIVE SCREW

NEEDLE

O-RING

COLOR RINGS

FLOW CONTROL VALVE

Figure 21

knob is turned clockwise. The color coded rings provide a convenient reference.

The lock screw is tightened when the flow rate defined by the operating window of opportunity is achieved. This establishes the desired plunger velocity.

Limit switches are the sensors and act as the eyes of the die casting machine. They read the position of the machine component movement and allow the activation and/or deactivation of the solenoid valves to change machine conditions at a predetermined position. These switches are mechanically operated and may perform a function at the time they are released, operated, or both.

These switches are simple in design, consisting of a housing containing a contact block, and an actuator head assembly. The actuator is made up of a lever arm, shaft, and a return spring. The plunger is the mechanism that connects the contact block and the actuator head.

Two sets of contacts are mounted to the contact block. One set is normally closed and the other is normally open.

The operating arm reverses their condition when it is moved. The details are given in Fig. 22.

In a "logical sequence" machine, the position of many components must be known at the same time.

Several components could be moving at the same time, and these movements must be coordinated to keep from damaging the machine or the product. An example of a limit switch arrangement and the functions performed is described in Figs. 23 and 24.

Limit switch number 1 (LS1), die retract stop, is adjustable and actuated when on the die lock stroke and *stops the retract movement* when actuated.

LS9, ejector core retract, is adjustable and *signals the ejector core to retract* when operated during die opening. It must be operated prior to LS7 when both are utilized.

LS15, cushion die release is adjustable and is operated on the die retract stroke. The dies should be open at least 1 in.

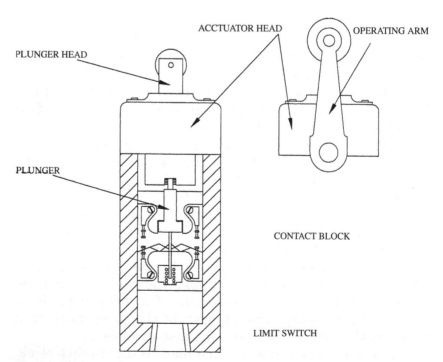

Figure 22

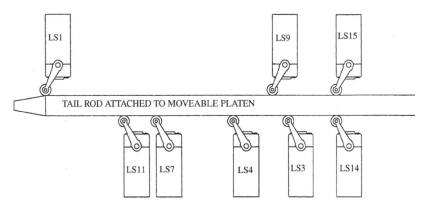

Figure 23

before operating because it *allows the die lock to go to fast retract velocity.*

LS14, cushion die forward, is adjustable and released on the forward travel of the die lock. When released, it *de-energizes the regenerative assist solenoid.*

LS2, die locked, is not adjustable, and is released when the *machine toggle linkage is in full locked condition.*

LS102, die locked, also functions when the machine is in full locked position.

LS3, shot retract, is adjustable and operated on the die retract stroke. When operated, the slow shot forward

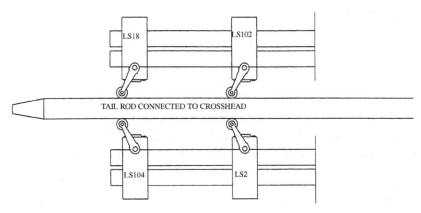

Figure 24

solenoid will be de-energized. *This allows the shot to retract.*

   LS4, low pressure on, is adjustable and is positioned to be released when the die lock starts forward. An obstruction will cause the die lock pressure to actuate this switch, *stopping the die lock forward movement.*

   LS7, mid-die stop, is adjustable and operated during die retract. *It provides a pause in the die retract movement to pull the ejector core before the casting is ejected.*

   LS11, ejection, is adjustable and operated during die retract *allowing the ejector circuit to be set up.* Ejection will start when the position of LS1 is reached.

   LS18, high pressure accumulator, is adjustable and released on forward travel of the die lock. When released, *the die accumulator close solenoid will energize.*

   LS104, low pressure off, is adjustable and released on the forward travel of the die lock. Used in conjunction with LS4, this switch should be *positioned to be released when the die faces touch.* At this point, low pressure close is no longer active.

   All the limit switches should be maintained properly in order that no false signals are given to the machine that can result in injury or damage to the equipment. Any time that a limit switch is defeated or held out, the employee or the equipment is in jeopardy.

   In addition to limit switches there may be proximity switches or magnetic strips on various machine components. While the limit switches are mechanical devices, the proximity switches and the magnetic strips are wired into a computer system.

   There are two types of ejection systems, for removing the shot from the ejector die half. One is the "bump bar" system that utilizes long bumper bars that actually contact a solid ejector plate when the dies are completely open. This pushes the shot forward into the die height area for removal. The other system is hydraulic with one or more cylinders that push the ejector plate forward to a point of removal from the die height area.

   The difference between the two methods is that the bumper pins stop the movement of the ejector plate so the shot stops while the die keeps moving and a hydraulic cylinder pushes the ejector plate forward to move the shot away from the die. A section is cut through a die to illustrate both ejection methods in Fig. 25.

The moving components of the ejector system should always be guarded for the protection of the operator. Many pinch points make this a very hazardous area.

The shot end of the machine injects the liquid casting alloy into the metal feed system and finally into the die cavity. There are two different injection systems: the "hot chamber" and the "cold chamber."

Hot-chamber machines are used primarily for casting zinc and magnesium, but also are used for other low temperature alloys such as lead. The alloys cannot be invasive or corrosive to the steel pot or gooseneck.

In this process, the injection mechanism is called the gooseneck. It is always immersed in the liquid metal bath of the holding furnace or crucible. There is a port in the gooseneck just down stream from the plunger tip so that liquid metal will fill the shot sleeve by the force of gravity. Note that the shot plunger is vertical which moves forward (down) into the shot sleeve to force the molten alloy up through the nozzle and onto a sprue or spreader pin that directs the metal stream into the runner and finally into the die cavity through the gate.

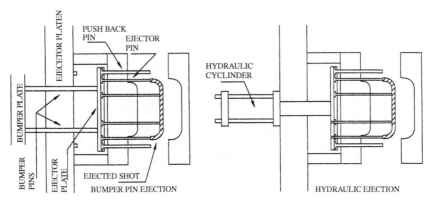

Figure 25

HOT CHAMBER

Die casting is a thermal process, so the hot chamber is more thermally efficient because the metal is never ladled and subjected to that loss of heat. Fig. 26 illustrates this condition. This process requires less pressure and produces castings at a faster production rate, about twice that of the cold-chamber method.

Typical operating pressure applied to the metal ranges from 1500 to 5000 psi in the hot-chamber process, while cold-chamber machines operate from 3000 to 15000 psi. However, the actual pressure used for each type of casting depends upon the quality requirements of the part and the design of the die, as well as the casting alloy.

COLD CHAMBER

Cold-chamber machines are designed with a metal injection system that is not immersed in the liquid metal bath because it is used primarily to cast aluminum alloys. Liquid aluminum acts as a solvent for iron and thus would rapidly dissolve the steel components in the cold-chamber system.

The horizontal shot sleeve, shown in Fig. 27, is normally in the horizontal position into which metal is ladled precisely, either manually or automatically, to minimize splashing. As the plunger advances, it seals off the pouring well and forces the metal into the die, first at slow speed, and then at high speed and pressure.

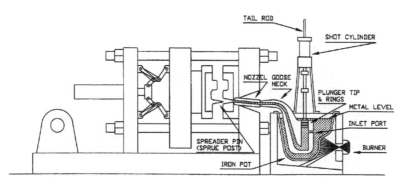

Figure 26

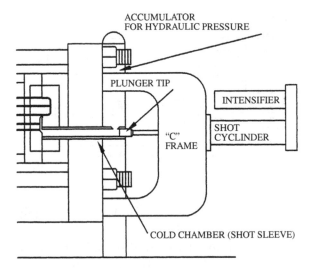

Figure 27

When the shot has solidified, the die opening and the plunger forcing the biscuit from the shot sleeve are synchronized with the ejection and the shot is removed to clear the machine for the next cycle. The production rate is considerably slower because of the longer solidification time for higher temperature alloys and the additional time needed for ladling.

The machine must be level so that it will stay in place during production and provide a reliable platform and reference for the die casting die. The basic advice that is offered here will prevent the powerful machine from distorting the more delicate dies that are precisely constructed to maintain the dimensional tolerances of the near net shape to be cast.

The machine should never be placed on the floor without a thicker reinforced concrete foundation installed directly under the machine location in a manner similar to a building footing. In addition, steel plates must be located under the leveling screws on the base. Figure 28 may be used for reference for the following recommendations.

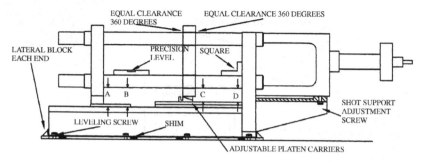

Figure 28

The purpose of the *screws* is to adjust the level at 24 inch. intervals all around the base rails, using a precision level. It is important that the machine be level within 0.0005 in. per foot. For transverse levels, a parallel bar can be placed across the base plate. This procedure should be followed at the edges and mid-point of the stationary platen and the back platen.

Levels are also important along and across the tie bars. The tie bars need to be square with the stationary platen. On machines equipped with a shot support screw, this square-ness may be accomplished by adjusting the screw. In the event that the machine has no shot end adjustment, shims are acceptable, but there must be contact between the bottom of the platen and the base rail.

Accurate measurement at A = B and C = D between the bottom tie bars and the base rail must be within a tolerance of 0.005 in. so as not to introduce mechanical dimensional discrepancies into the vulnerable casting process.

Both ends of the machine need to be securely blocked to prevent movement. This movement of unsecured machines is called "walking."

Equal clearance around the entire circumference of the tie bars and the bushings in the moving platen is accomplished by adjusting the carriers under this plate. This clearance should be the same at both minimum and maximum die height positions.

Preventive maintenance is an essential element that must be recognized by the management if predictable results are desired. The die casting industry has a reputation for subjecting casting equipment to very severe operating conditions. Of course, these conditions vary from plant to plant, but are a reflection of the management function. These suggestions are aimed at improved up time by planning maintenance before a break down forces the machine down.

Though improvements have been made during the last decade, uptime of North American die casters does not compare favorably with that of competitors in Europe and the Asian basin. Die casting is a high fixed-cost process and most of these costs relate to the casting machine. Runtime in the ninety percentile is the level that identifies the survivors. Critical areas are outlined here that should be given strict attention with routine monitoring on a regular and recorded basis.

Hydraulic fluid is vital to the operation of the casting machine and has to be properly cared for. Most of the fire resistant fluids in use today have a detergent action and are therefore prone to foam and entrain air. Lubricating qualities are also lacking for environmental reasons, are more susceptible to changes in viscosity, and display a high specific gravity.

These qualities all contribute to malfunction of the hydraulic system unless they are conscientiously controlled. Thus, periodic testing by the fluid manufacturer is basic to the uptime performance of the machine.

The fluid has to be replaced immediately if tests indicate contamination or a change in viscosity. New clean filtered fluid should be used, but only after the machine reservoir and hydraulic system have been thoroughly drained and flushed. Continuous or scheduled filtration will greatly enhance machine performance.

The temperature of the fluid should be in the range of 90–115°F or that specified by the manufacturer, if different for water glycol, the material of choice. It the temperature is lower than this range, it becomes thicker and more difficult to pump. This condition puts too heavy a load on the pumps and can affect machine performance. Higher temperature causes the fluid to become thin and lose lubricity. Excess wear on pumps and valves results. More leakage past value spools also adds to the aggravation. When water glycol overheats, water evaporates faster so that the fluid breaks down.

The level of fluid in the reservoir must be maintained as specified by the machine manufacturer since the pump will take in air causing cavitation. Fluid exits the return lines at a high velocity and when they are not sufficiently submerged, the surface is disrupted causing air entrapment.

The color of the fluid can signal trouble. Entrapped air gives the fluid a "milky" appearance. When viewing it through the level gage, a foreign material would float on top in all probability if it is mineral based.

Hydraulic system maintenance, though closely associated with the fluid, is another critical factor in machine uptime. Pumps and valves wear even under the best of conditions. Minimizing extensive damage requires regular inspection and replacement of certain parts or assemblies. Frequency depends upon the degree of attention given to the above points.

The reservoir must be cleaned after the failure of any hydraulic component, especially pumps. This requires draining of the old fluid. The covers are removed and each compartment is flushed with a high pressure jet of kerosene or water. Lint free cotton cloths are used for wiping down the tanks after flushing.

All pump suction lines, filters, strainers, and magnets also have to be thoroughly cleaned. Whether new or used fluid is used to fill the reservoir, it must be passed through a 15 μm filter.

The heat exchanger is often neglected and should be removed and cleaned annually. Hydraulic components can be damaged by tube fouling in the exchanger, which causes the temperature of the hydraulic fluid to rise.

Cleaning requires that the shell side be plugged and immersing or flushing it with Oakite® or a similar cleaning solution in a heated tank. New gaskets should be installed when the unit is reassembled.

Raw water can be highly corrosive, but can be treated to prevent build up of scale in an action similar to that which occurs in die cooling channels. Salt water will cause galvanic corrosion, which can be prevented by the insertion of zinc sticks in the heat exchanger.

Leaks occur at threaded connections either because the pipe cracks, usually at the root of the thread closest to the fitting, or the thread fit is loose. This is a real problem to the die casting industry because it diminishes the quality of the workplace as well as wastes the leaked fluid. Cracks are caused by vibration of the piping system. The obvious fix is to add more supports to the piping. Loose threads call for replacement of the pipe or fitting.

On the surface, this looks like such a simple problem, but when the great quantity of connections on a single casting machine is multiplied by the number of machines in the plant, the situation becomes overwhelming. One way to deal with it is to constantly repair the leaks as they occur. Of course, proper installation of connections is the best approach. Many times repairs are slow because the machine will function without them, and production is always the highest priority.

Lubrication is absolutely necessary due to the extreme forces placed upon all moving parts, especially the toggle linkage. A clean film of lubricant has to be maintained at all times. Daily monitoring of all lube lines, fittings, reservoirs, motors, couplings, etc. is essential.

The electrical system is vulnerable to shorts. Control panel doors have seals to prevent ambient air from entering. Cooling air is filtered because dirty air can cause a short. Many violations can be observed at this point.

Junction and terminal boxes should be sealed with covers at all times to keep out moisture and dirt.

The mounting of limit switches must be checked to ensure that the switch is tightly and securely mounted and in the proper location. Loose switches can be activated at the wrong time and position. Warning and indicator lights must be checked to ensure that they work at the right time and are visible to all people involved.

Batteries in all electronic devices must be regularly replaced because they are used to hold the program when the power is off.

To summarize this chapter, it is obvious that the casting machine is not a simple device. There are thousands of moving parts to keep track of as well as several technical disciplines. The disciplines are: mechanical, hydraulic, electrical, and structural. Special skills are required in every die casting facility to function effectively. This is the human equation in which the demand for these skills is usually greater than the supply.

The reader is urged not to over simplify the casting machine, but the functions of the machine can be grasped more easily than the details. A casting machine is really a clamp on one end to hold the dies closed, and a pump to supply superheated liquid metal to the die on the other end.

The clamping force available through the tie bars determines the projected casting area that the machine is capable of producing.

The pumping capacity of the machine is determined by the ratio of the area of the shot cylinder to the plunger tip in addition to the range of plunger velocities during the slow and fast shot phases.

The proper utilization of these features drives the quality of the near net shape produced.

4

Casting Metallurgy

Die castings are produced from alloys composed of two or more metals. The predominant metal is usually either aluminum, magnesium, zinc, or, in some cases, lead or tin.

In each alloy system, the predominant metal is called the base metal if it exceeds 50%, and is expressed first when naming a particular alloy. Those constituents that are present in an alloy are named after the base metal to describe the system. Some more common *pure metals and alloy systems* are:

Aluminum-base alloys

380 and 383 aluminum–silicon–copper (Al–Si–Cu) system.

413 aluminum–silicon (Al–Si) system.

390 aluminum–silicon.

360 aluminum–silicon–magnesium (Al–Si–Mg) system.

518 aluminum–magnesium (Al–Mg) system.

Aluminum alloys are sold in ingot (primary or secondary) or liquid (hot) form.

Binary alloys are specified as 2xx, ternary alloys are identified as 3xx, while the eutectic is referenced as 4xx.

Alloy 380 and 383 are hypoeutectic, 413 is eutectic, and 390 is hypereutectic.

Magnesium-base alloys

AZ91 D magnesium–aluminum–zinc (Mg–Al–Zn) system.

AM60 B magnesium–aluminum.

Zinc-base alloys

No. 3 and No. 7 zinc–aluminum (Zn–Al) system.

No. 2 and No. 5 zinc–aluminum–copper (Zn–Al–Cu) system.

ZA8, ZA12, and ZA 27 (Zn–Al) system.

Die castings must satisfy a wide range of requirements from cosmetic to structural and the performance depends upon the properties of the chemistry of the alloy from which the casting is made. These properties are a function of the alloy constituents, contaminants, solidification patterns, and treatments performed after casting.

Each pure metal has a characteristic cooling pattern as it transforms from the liquid state to the solid state, and the reverse when it is melted. As time passes, heat is removed, and the temperature drops. While the metal solidifies, however, the temperature does not change, even though heat is being removed. Metals that lend themselves to rapid solidification and still maintain desirable physical properties are most commonly used in the die casting process. For these pure metals, this phenomenon is graphically illustrated in the time, temperature, transformation (TTT) chart in Fig. 1.

The flat portion of the curves describes the phase when the metals give up the heat of fusion, which is called the

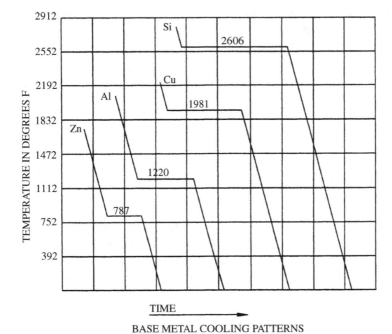

TIME

BASE METAL COOLING PATTERNS

Figure 1

eutectic of the metal. The perfect flat condition shown in the chart will not be found in the real world where base metals are alloyed with other elements. A typical TTT curve will describe the flat line of a die casting alloy at a down sloping angle because the perfect metallurgical condition is compromised.

Two critical temperatures for these metals are the *melting point* and the *heat of fusion*, and Table 1 will serve as a convenient reference.

These data will help to visualize the thermal behavior of the base metals used in die casting. Note that the steep pitch of the TTT curves describes rapid solidification. The table tells us that it takes more than four times as long for aluminum to solidify than for zinc because the heat of fusion is greater.

It must be understood, however, that this is very basic and fundamental casting alloy information that is presented in a simple form here for purposes of brevity. *The rapid*

Table 1 Melting point and Heat of Fusion of Some Common Die
Casting Alloy Base Metals

Metal	Melting point (°F)	Heat of fusion	
		cal/g	BTU/lb
Aluminum	1220.4	94.6	170.0
Copper (brass)	1981.4	50.6	91.1
Magnesium	1202.0	89.0	160.0
Silicon	2605.0	337.0	607.0
Zinc	787.03	24.09	43.36

solidification rate of all die casting alloys distinguishes high
pressure die casting from the other foundry processes. When
the alloys change from the liquid to the solid state, the quick
freezing rate is important to crystallization. Thus, die castings have a fine grain size, dense structure, and superior
mechanical properties that make this process superior to
other casting processes.

The structural properties of the castings produced are
affected by the environment in which solidification occurs. It
is defined by the combination of metals in the casting alloys
and their atomic composition. During solidification, atoms
form crystals that become relatively dormant. Atoms of each
metal are oriented into specific relationships within the crystals. The arrangement of atoms for each metal displays certain identifiable patterns. This phenomenon is called a
lattice shape, which normally defines the properties of the
casting alloy.

Atomic movement into and out of the crystal structures
occurs, even in the solid state. This accounts for the solid solubility of each element. The faster freezing rates experienced
in high pressure die casting diminish this movement. This
explains the finer and more dense grain structures that
enhance the mechanical properties.

During solidification, one metal governs the behavior of
the crystal lattice structure. This is why each casting alloy
demonstrates a specific freezing range. During this time, both
liquid and solid phases exist. The actual state of the alloy can

be described as slushy during this period between the liquidus and solidus temperatures. When the eutectic occurs prior to the end of cavity fill, internal defects can be expected.

Initially, crystals are formed that are composed of the base metal (aluminum, magnesium, zinc, etc.). The crystal lattice is repeated until solidification is finished. Then, the individual crystals contact each other at the grain boundaries, which establishes the grain structure. During solidification, crystal growth progresses and forms dendrite arms. The freezing rate determines the size and spacing of the dendritic structure. This is not critical in die casting because of the rapid freezing. As a matter of fact, the distribution is similar to castings produced by other processes that are heat treated to the T4 level at solution temper. For this reason, high pressure die castings are rarely heat treated. In today's competitive environment, the array of alloys has expanded and the application of heat treatment is becoming more prevalent.

The crystals that form when liquid metals or alloys solidify are also called grains. The crystal or grain is a three-dimensional pattern of atoms. For a particular alloy, the configuration is always consistent in the pattern of one of 14 atomic configurations in the lattice. This configuration is referred to as a *structure* that repeats for any element or, in the case of die cast materials, any alloy. The smallest unit in the structure is the cell that is constantly copied as the crystal develops.

Since the rate of growth depends on time and temperature, the crystals are not always uniform. Each crystal forms on a nucleolus so the availability of nuclei determines the quantity of crystalline growth. This takes place in liquid alloy in which the crystals are dissolved. The growth is stopped by contact with adjoining crystals. Atoms that are dissimilar to those in the lattice also affect the size and shape of the crystals.

Solid solutions are the result of dissimilar atoms entering the lattice of another metal. The lattice that is thus formed determines the properties of the alloy. The rate of heat extraction establishes the crystal or grain size. The rapid solidification that occurs in the die casting process increases the quantity of crystals and reduces the size of each. The face centered cubic crystal illustrated in Fig. 2 is frequently found in many metals. The atoms at the center and each corner of all the faces are typical of both aluminum and copper.

The atoms are surrounded by loosely held electrons which, when shed, form Al–Cu ions. Think of these free electrons as a gas around the ions. The charge balance between the positive atoms and the negative electron gas develops a metallic bond. This bond is also possible between atoms of dissimilar metals with different structures. The distance between the centers of the atoms at the corners of the edge of the unit cell defines the lattice parameter.

There are four planes and 12 directions in the face centered cubic lattice. This is the greatest concentration of atoms

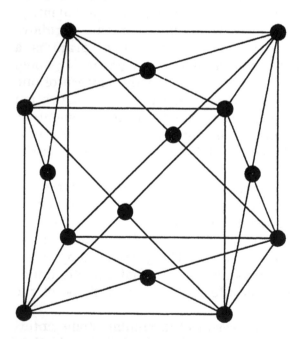

THE FACE CENTERED CUBIC LATTICE IS
TYPICAL OF SOME DIE CAST METALS
LIKE ALUMINUM AND COPPER

Figure 2

and symmetry found in any lattice structure. This is why copper and aluminum are so ductile. By the same token, the loosely held electrons in the electron gas make these metals good electric and heat conductors.

Tungsten, molybdenum, and iron are used in die materials because of their high strength and reasonable ductility. These metals are defined by the body-centered cubic lattice. A unit cell with an atom at each corner and one in the center of the cube is illustrated in Fig. 3.

The other major die cast metals are magnesium and zinc. They crystallize in the close-packed hexagonal lattice. The close-packed patterns of atoms are similar to the face centered cubic lattice. The top and bottom faces display a hexagonal grouping where one atom is surrounded by six others. The concentration of atoms in these planes is parallel. There are

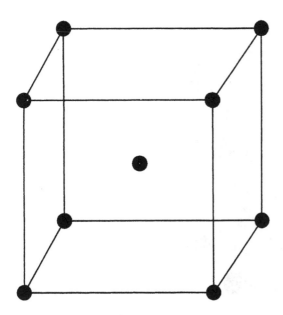

A BODY CENTERED LATTICE
DESCRIBES THE CRYSTALIZATION
OF TUNGSTEN, MOLYBDENUM, AND
IRON

Figure 3

a total of three of these close-packed planes described by the lattice in Fig. 4.

Plastic deformation can therefore be considered to proceed along specific planes and directions. However, it is more limited than the face centered cubic structure with four planes in 12 directions.

Being hexagonal elements, zinc and magnesium display about the same plasticity as face centered cubic elements but are more ductile than body centered cubic metals. Movement between planes is enhanced by additional slip planes by a mechanism called *twinning*.

Metals that are more brittle crystallize with less symmetry than both face-centered and body-centered cubic configurations. They do not display close-packed planes and the slip directions are not well defined. These metals are therefore not widely used.

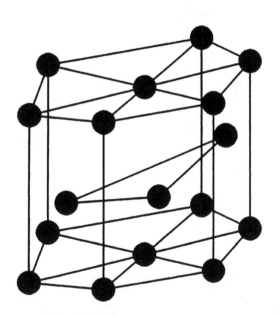

MAGNESIUM AND ZINC SOLIDIFY IN
THE CLOSE PACKED HEXAGONAL
LATTICE STRUCTURE

Figure 4

Several different lattices result when groups of metals are melted together and are then allowed to solidify. The effect of temperature on the intermediate phases and mutual solubility defines the type of lattice that is thus formed when the elements become solids. The different types of structures can be graphically described by equilibrium diagrams.

There is some solid solubility between any two metals in an alloy even though the solubility may vary greatly. A binary system occurs when the solid solubility is small. Unless the metal that acts as the solvent can exist in more than two crystalline forms, the lattice will continue until the melting point is reached.

Usually the crystal lattice of the solvent is maintained in a solid solution even though the separate metals have different crystal structures. This condition is referred to as a *phase*. There are two types of solid solutions.

In one, the solute atoms are substituted for atoms in the solvent structure. This substitution changes the size of the lattice parameters in the solvent cell. The extent depends upon the size of the atom that is substituted.

In the other solid solution, the atoms of the solute move into the spaces between the atoms in the solvent. Thus, they become part of the structure of solvent atoms. This scenario occurs infrequently because the solute atom must be much smaller than the those in the solvent. The austenite phase that results from carbon that is alloyed into the iron in the premium grade H13 and P20 steels used in casting die cavity inserts is an example of this type.

It must be emphasized that this whole discourse can only be observed micrographically and is therefore not obvious to most of us who must work with metals on the shop floor. Primary crystals freeze out of liquid alloys and continue to grow in a tree shaped arrangement. This configuration is usually referred to as dendritic and occurs as long as the primary element is available. Most die casting alloys solidify in both liquid and solid phases. Rapid solidification restricts the dendritic growth to form the fine dense grain structure.

Dendrites form when casting alloys solidify so it is appropriate to discuss them and the grain structures. "Dendrite" derives from the Greek word for tree, which is the shape taken when they group together in a grain. A grain is a family of dendrites that originate from the same nucleolus.

The finger shape of dendrites drives the latent heat of fusion away from the liquid–solid interface. A more familiar but similar heat transfer takes place when the fingers of your hand are exposed to cold. Gloves retard some of the heat escape, but mittens work better because there are no fingers. During the rapid solidification that occurs in die casting, the formation of dendrites takes place on a schedule defined by the chemical composition of the alloy. It is gradual even though also rapid as shown in Fig. 5.

Metallurgically speaking dendrite fingers are called arms. The intricate network of arms inhibits free movement of the remaining liquid alloy during solidification so that the microscopic spaces formed between the arms are starved of the liquid necessary to make up for solidification shrinkage. These voids are what is known as microporosity as depicted in Fig. 6.

Two dendrites are illustrated that have come together and the microporosity is described by the crosshatched area. Eutectic silicon is also found between the arms of aluminum alloys.

Understanding liquid and solid starts with the TTT chart that presents the thermal behavior of pure base metals. If the temperature reaches a level in which the metal is fully liquid, a point referred to as the liquidus has been achieved. However, die casting alloys are not pure metals as they are usually a combination of two (binary) or three (ternary) base metals. The behavior of a specific alloy is determined by this combination.

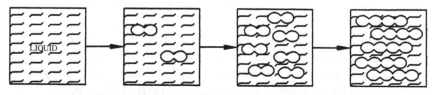

PROGRESSION OF DENDRITE FORMATION DURING SOLIDIFICATION

Figure 5

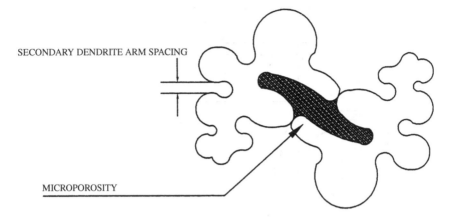

SECONDARY DENDRITE ARM SPACING

MICROPOROSITY

Figure 6

The latent heat of fusion, sometimes referred to as the heat of transformation occurs when the melting point or freezing temperature is reached. The TTT chart defines a temperature that remains constant as solidification continues, even though heat energy is lost. The total energy given up during freezing is defined by the length of the flat part of the cooling curve. The transformation from the liquid to the solid state occurs because of this energy exchange. The amount of heat is specific and defined in terms of BTUs per pound or calories per gram. When defined in this manner, it is called the *latent heat of fusion*. The equilibrium diagram in Fig. 7 provides a graphic explanation.

The eutectic is the lowest melting point of a metal in an alloy system. Therefore, the flat part of the TTT curves is called *eutectic arrest*. Aluminum alloy A13 is sometimes referred to as the eutectic alloy because of the effect of the 12.6% silicon content upon the liquidus temperature. Since the 380 binary aluminum alloy is in such common use, an equilibrium diagram for this aluminum silicon alloy is offered (Fig. 7). Note that the fluidity varies about the eutectic line for silicon. In addition to freezing characteristics similar to pure metals, eutectic alloys in the solid state are homogeneous mixtures, under conditions of equilibrium, of the combining metals. The metals may also be described as isothermal

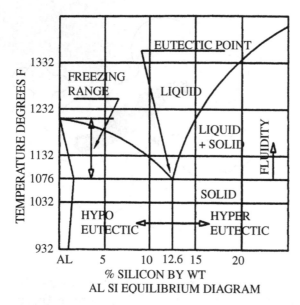

Figure 7

reversible reactions so that the liquid combination solidifies into two intimately mixed solids upon cooling.

Chemical composition in die castings is alloyed into secondary aluminum alloys that are generated from scrap rather than from primary material that is refined from the bauxite ore that is the original source of aluminum.

The composition of the most commonly used die casting alloys is expressed as a percentage by weight in Tables 2–5.

Zinc is normally supplied to the die caster in the pure

Table 2   Chemical Composition of the Three Grades

	Composition % by weight			
Grade	Lead (max.)	Iron (max.)	Cadmium (max.)	Zinc (min.)
Special high grade (SHG)	0.003	0.003	0.003	99.99
High grade (HGZ)	0.03	0.02	0.02	99.90
Prime western (PW)	1.40	0.05	0.20	98.00

Table 3 Chemical Composition of Two Major Alloys

Designation	CU	AL	MG	FE (max.)	PB (max.)	CD (max.)	SN (max.)	NI	ZN
NO. 3	0.25 max	3.5–4.3	0.02–0.05	0.10	0.005	0.004	0.003	–	REM
NO. 5	0.75–1.25	3.5–4.3	0.03–0.08	0.10	0.005	0.004	0.003	–	REM
ZAMAK 7	0.25 max	3.5–4	0.005–0.02	0.075	0.003	0.002	0.001	0.005–0.02	REM

Table 4 Chemical Composition of Aluminium Alloys by Weight in Castings

Commercial designation	Copper	Iron	Silicon	Magnesium	Manganese	Zinc	Nickel	Tin	Others
360	0.6	2.0	9.0–10.0	0.40–0.60	0.35	0.50	0.50	0.15	0.20
380	3.0–4.0	2.0	7.5–9.5	0.10	0.50	3.0	0.50	0.35	0.50
A380	3.0–4.0	1.3	7.5–9.5	0.10	0.50	3.0	0.50	0.35	0.50
383	2.0–3.0	1.3	9.5–11.5	0.10	0.50	3.0	0.30	0.15	0.50
384	3.0–4.5	1.3	10.5–12.0	0.10	0.50	3.0	0.50	0.30	0.50
390	4.0–5.0	1.3	16.0–18.0	0.45–0.65	0.10	0.10	–	–	0.20
A13	0.6	1.3	11.0–13.0	0.10	0.35	0.50	0.50	0.15	0.25
43	0.6	0.8	4.5–6.0	0.05	0.50	0.50	–	–	0.35
218	0.25	1.8	0.35	7.5–8.	0.35	0.15	0.15	0.15	0.25

Table 5 Chemical Composition of the Major Magnesium Alloys Used in Die Casting

| Designation | Aluminum | Zinc | Chemical composition | | | | | |
			Manganese (min.)	Silicon (max.)	Copper (max.)	Nickel (max.)	Iron	Others
AZ91B	8.5–9.5	0.45–0.90	0.15	0.20	0.25	0.01	NO SPEC	0.30
AZ91D	8.5–9.5	0.45–0.90	0.15[a]	0.02	0.015	0.001	0.005[a]	0.01
AM60B	5.5–6.5	0.22 max	0.24–0.6[a]	0.10	0.010	0.002	0.005[a]	0.02

[a]If either the minimum MN limit or the maximum FE limit is not met, then the FE/MN ratio shall not exceed 0.01 and 0.021, respectively.

slab form and is then economically alloyed in house. Table 2 describes the chemical composition of the three grades.

The chemical composition of the two major alloys are defined in Table 3.

The chemical composition of aluminum alloys by weight in castings is given in Table 4.

Note that ingot or liquid metal purchased by the die casting firm is held to a tighter iron specification because of the solubility of iron in aluminum.

The chemical composition of the major magnesium alloys used in die casting is covered in Table 5. Though not stated in Table 5, magnesium constitutes the remainder of the chemical composition.

Aluminum is not used without alloying for any purpose except for electrical motor rotors because of its low strength and hardness as well as its poor machinability. Other elements are therefore added to improve upon these properties. The elements most commonly alloyed with aluminum are copper, silicon, and magnesium. To a lesser extent, manganese, iron, zinc, and nickel are alloyed. In general, the addition of elements to aluminum is limited to approximately 15%. Beyond this point, alloys become increasingly brittle, which takes away from their engineering value.

Copper improves the strength and hardness progressively until it reaches a level of about 4%. Above this point, the alloy becomes too brittle. It greatly enhances machinability and also improves properties at elevated temperatures. It lowers corrosion resistance but increases fluidity.

At 4%, copper increases the tendency for hot cracking, but further additions decrease the incidence of hot cracking.

Silicon is an important addition to aluminum alloys since the casting characteristics are greatly enhanced. There is a progressive improvement in fluidity with a reduction in hot cracking. Up to the eutectic point of 12.6%, the incidence of solidification shrinkage decreases, making it easier to produce castings free of shrinkage and cracks. This condition suggests that the Al–Si system alloy is the choice for pressure tight castings. Care must be exercised, however, as this is a premium priced alloy.

The trade off for an increase in strength and hardness is a commensurate decrease in ductility. These properties, however, are improved by the rapid solidification.

Magnesium produces a gradual increase in strength up to 6%, although hardness is not effected by magnesium until the 10% level is reached. Therefore, the binary Al–Mg aluminum systems have excellent mechanical properties, resist corrosion, and are very machinable. The impact resistance is good, as is ductility, and they maintain these good properties at elevated temperatures.

Why then are these alloys not used more? The fluidity is so poor that castibility becomes a real problem. The solidification range for these alloys is also very narrow, so that premature freezing during cavity fill must be carefully handled through thorough mathematical analysis.

Iron is a natural ingredient in aluminum alloys due its association with iron in bauxite ore and the aggressive affinity that iron has to go into solution with aluminum. Some metallurgists even go so far as to call aluminum the universal solvent. For this reason iron crucibles cannot be used to hold liquid aluminum, as the bath will eventually dissolve the pot.

Iron forms a eutectic with aluminum at 1.7% and it has a solidification point of 1211°F. Although iron is commonly considered an impurity, it performs a useful function as long as the content is below 1.7%. It increases strength and hardness and reduces the tendency for hot cracking. The limit in ingot or liquid alloys is 1%; iron up to 1.7% materially reduces soldering and is allowed in castings. In this writer's experience, secondary aluminum with iron as an allowed impurity is preferable to primary alloy because of the tendency toward higher iron content.

This tendency for iron pick up when the alloy comes into contact with steel dies and shot sleeves limits the number of passes a batch of aluminum can make before being resmelted. Iron content should not exceed the 1.5–2.0% level, provided that manganese and chromium are present to avoid large concentrations of Fe Al3 needles in the microstructure.

Manganese and chromium are beneficial in small quantities, but the propensity to sludge becomes a problem if the levels get out of control. Excessive sludging is the penalty for losing control and is to be avoided. It is a major contributor to melting loss of metal.

A *word about hypereutectic aluminum–silicon alloy 390* is appropriate even though a low tonnage is produced. The high silicon content of 390 alloy of 16–18% requires a melting point above the eutectic because the melting point of silicon is 2606°F. Therefore, the rules of casting this alloy are considerably different than for the other aluminum alloys.

The alloy was designed for the production of automobile engine blocks to replace the need for expensive iron cylinder bore inserts and to replace cast iron blocks to reduce total car weight. The technology consists of the alloy itself, compatible piston material, and a special cylinder bore finish after machining.

The hard primary silicon phase is abrasive to cutting tools so polycrystalline coated diamond tools must be used. It does, however, have very desirable machining characteristics. Built-up edge on the cutting tool tends to be less than when machining more conventional alloys, and chips are short and easy to handle. Desired surface finishes are readily achieved. The 390 alloy bore finishing concept calls for the primary silicon to stand slightly proud from the bore surface (Lee et al., 1991).

Sludging is generated by the wrong combination of FE–Mn–Cr, so an evaluation is necessary to monitor and control it. The formula for this calculation follows.

Iron, manganese, and chromium form complex intermetalic compounds in aluminum base alloys. These compounds possess extreme hardness and high melting points. These elements precipitate out of liquid solution because they have a higher specific gravity than the parent aluminum. Crystals may form at temperatures higher than the liquidus and may chemically combine into complex intermetallics. When this occurs, they acquire very high melting points and do not easily redissolve. They begin to coalesce and their higher specific gravity causes them to sink to the bottom of the melt where a sand or sludge is formed.

Manganese has proved to be a powerful agent in causing the formation of sludge; chromium, usually encountered in lower concentration in secondary aluminum alloys, has an even stronger influence. The sludge that forms is crystalline and sugary looking in appearance and may contain from 4% to 20% iron. Even small concentrations of sludge injected into the die cavity will cause casting and machining problems.

Several empirical sludging formulae have emerged which offer a reasonable guide for predicting whether a melt will be prone to the formation of sludge. Compliance with the formula, however, does not guarantee that sludge will not occur because the total factor may vary widely under different melting conditions. Also, some alloys are more or less sensitive to sludge formation than others.

The sludge formula usually used is:

$$\%Fe + 2x\ \%Mn + 3x\ \%Cr = 1.80\,max$$

Zinc-base alloys are also affected by alloying elements or impurities and will be examined here.

Iron is vulnerable to zinc and rapidly alloys with it, especially at temperatures above 850°F. A small addition of 0.25% minimum of aluminum reduces this tendency at normal operating temperatures. Most zinc die casting firms alloy Zn–Al systems from slab zinc; this addition of aluminum is called hardener.

Since the zinc alloy comes into contact with iron during the casting process, iron pick up can exceed the 0.10 specification limit. This combines with aluminum to form the intermetallic compound of $FeAl_3$, which is lighter than the alloy and floats to the surface and becomes the primary mechanism for the formation of dross or skimmings. It is, therefore, important to control the temperature of the melt at 850°F maximum so that excessive dross will not form on the surface of the holding crucible. An iron content of less than 0.02% may then be expected. Usual casting temperature of zinc alloys is 800°F, so the upper limit is not difficult to avoid.

Iron above the specification also causes cracking during subsequent secondary operations such as bending or staking.

Copper is considered an impurity but is not detrimental up to 1.25%. Excessive aging growth may be expected above this level. Strength and hardness are enhanced with the addition of copper.

Magnesium concentration that exceeds the specification limit will cause hot cracking and there will be a loss of fluidity. It does offset the effect of intergranular corrosion that will be discussed later.

Nickel within the solubility limit of 0.02% helps to neutralize those elements that cause intergranular corrosion.

Lead above 0.005% cannot be tolerated because it migrates to the grain boundaries of zinc alloy die castings.

Cadmium promotes drossing, hot shortness, and poor castability at levels above 0.1%.

Tin is not a natural impurity and enters the alloy as a contaminant from outside sources. Serious problems of intergranular corrosion and excessive aging growth occur.

Chromium above its solid solubility limit of 0.02% will form intermetallic compounds and float to the surface. The usual source is from remelting plated scrap. If chromium–aluminum compounds are formed, they may cause machining problems if they become entrapped in die castings.

Intergranular corrosion occurs when several impurities, but particularly lead, tin, and cadmium, exceed their limits and migrate directly to grain boundaries. These impurities are subject to chemical attack, especially in a warm, humid environment. When this happens, there is a swelling effect, followed by fracturing of the casting at the grain boundaries, and ultimately, the casting disintegrates.

Physical and mechanical properties define exactly how well an alloy can perform against a force that exposes the cast part to destruction. The chemistry of alloying other elements with the base metals of aluminum, magnesium, and zinc has been discussed in some detail that describes the role of each chemical element. Perhaps a reference back to the effects of the alloying elements when studying a particular property in the tables that follow will reveal the attribute that it brings to the quality of the casting produced. (Refer to Tables 6–8.)

This section describes the physical performance in quantitative terms that can be expected from each die casting alloy. It is important to point out that these are based upon as cast characteristics for individually die cast specimens, and not test portions cut from production die castings.

Young's modulus defines the stiffness or resistance to necking of the alloy. The slope of the curve up to the yield point implies stiffness and the elastic limit is described by the length of the horizontal portion of the curve. In Fig. 8, the top curve depicts the most brittle alloy when compared to the other two. *A typical stress strain curve*, is a method of depicting several different material properties. The length of the straight line portion of the curve indicates the strength. The point where each curve bends to the right is the yield point where the material starts to stretch when a tensile force is applied. Therefore, material A is the strongest.

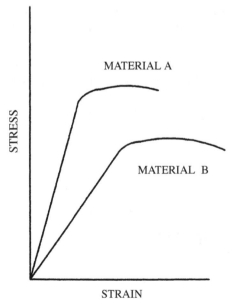

TYPICAL STRESS STRAIN CURVE

Figure 8

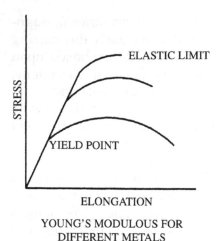

YOUNG'S MODULOUS FOR
DIFFERENT METALS

Figure 9

Another factor to be considered is toughness. In Fig. 9, the degree of toughness can be determined by the area under the curves. Thus, material B found in Fig. 8, is the toughest.

The slope of Young's modulous curve implies stiffness which is desirable up to a point but the material defined by the top curve is too brittle. The elastic limit is too short. A phenomenon referred to as necking is used to visually define the approach to the elastic limit in test specimens. When tensile forces are applied at both ends, the region at the center narrows just before the material fractures as illustrated by Fig. 10.

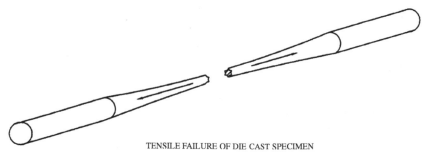

TENSILE FAILURE OF DIE CAST SPECIMEN

Figure 10

Table 6 Properties of Aluminium Alloys

Commercial designation	A360	A380	383	384	390	A13	43	218
Mechanical property								
Ultimate tensile strength (ksi)	46	47	45	48	46	42	33	45
Yield strength (ksi)	24	23	22	24	36	19	14	28
Elongation (% in 2 in.)	3.5	3.5	3.5	2.5	<1	3.5	9.0	5.0
Hardness (BHN)	75	80	75	85	120	80	65	80
Shear strength (ksi)	26	27	–	29	–	25	19	29
Impact strength (ft lb)	–	–	3	–	–	–	–	7
Fatigue strength (ksi)	18	20	21	20	20	19	17	20
Latent heat of fusion (btu/lb)	168.96	168.96	168.96	168.96	–	–	–	–
Young's modulus (10,000,000 psi)	10.3	10.3	10.3	–	11.8	–	10.3	–
Physical property								
Density (lb/cu in.)	0.095	0.098	0.099	0.102	0.098	0.096	0.097	0.093
Melting range (°F)	1035–1105	1000–1100	960–1080	960–1080	950–1200	1065–1080	1065–1170	995–1150
Specific Heat (btu/lb/°F)	0.230	0.230	0.230	–	–	0.230	0.230	0.230
Coefficient of thermal expansion (u in./in./°F $\times 10^{-6}$)	11.6	12.1	11.7	11.6	10.0	11.9	12.2	13.4
Electrical conductivity (% IACS)	29	23	23	22	27	31	37	24
Poisson's Ratio (mm/m)	0.33	0.33	0.33	–	–	–	0.33	–

Table 7 Properties of Magnesium Alloys

Commercial designation	AZ91D	AM60B	AM50A	AM50B
Mechanical property				
Ultimate tensile strength (ksi)	34	32		32
Yield strength (ksi)	23	19		18
Elongation (% in 2 in.)	3	6–8		6–10
Hardness (BHN)	75	62		57
Shear strength (ksi)	20	n/a		n/a
Impact strength (ft lb)	1.6	4.5		7.0
Fatigue strength (ksi)	10	10		10
Latent heat of fusion (btu/lb)	160	160		160
Young's modulus (10,000,000 psi)	6.5	6.5		6.5

Commercial designation	AZ91D	AM60B	AM50A	AM50B
Physical property				
Density (lb/cu in.)	0.066	0.065	0.064	
Melting range (°F)	875–1105	1005–1140	1010–1150	
Specific heat (btu/lb/°F)	0.25	0.25	0.25	
Coefficient of thermal expansion (u in/in./ F × 10^{-6})	13.8	14.2	14.4	
Thermal conductivity (BTU/sq. ft/hr/°F)	41.8	36	36	
Electrical conductivity (% IACS)	35.8	31.8	31.8	
Poisson's ratio (mm/m)	0.35	0.35	0.35	

Table 8 Properties of Zinc Alloys. Note the exceptional properties displayed by the ZA alloys.

Commercial designation	No. 3	No. 5	Z-A8	Z-A27
Mechanical property				
Ultimate tensile strength (ksi)	41	48	54	62
Yield strength (ksi)	–	–	41–43	52–55
Compressive yield strength (ksi)	60	87	37	52
Elongation (% in 2 in.)	10	7	6–10	2.0–3.5
Hardness (BHN)	82	91	100–106	116–122
Shear strength (ksi)	31	38	40	47
Impact strength (ft lb)	43	48	24–35	7–12
Fatigue strength (ksi)	6.9	8.2	15	21
Young's modulus (10,000,000 psi)	–	–	12.4	11.3
Physical property				
Density (lb/cu in.)	0.24	0.24	0.227	0.181
Melting range (°F)	718–728	717–727	707–759	708–903
Specific heat (btu/lb/°F)	0.10	0.10	0.104	0.125
Coefficient of thermal expansion (u in./in./ °F $\times 10^{-6}$)	15.2	15.2	12.9	14.4
Electrical conductivity (% IACS)	27.0	26.0	27.7	29.7
Poisson's ratio (mm/m)	0.030	0.030	0.030	0.030

The die casting industry is quantified by weight even though most calculations and measurements are by volume. Metal is purchased by the pound or ton; sales are announced in dollars, tons, or pounds; and melting capacity is designed for pounds per hour, etc. This is in contrast to technical units of measure that usually describe volume.

A reduction in volume occurs when a metal cools from the liquid state to the solid state which is called shrinkage (Doehler, 1951). This phenomenon applies to the dimensions of all casting processes. It is more pronounced in die casting because of the rapid solidification that occurs during this process.

The amount of shrinkage of a given alloy depends upon its chemical composition, but other conditions also may have an effect, i.e., the injection temperature, the temperature of the die surface, the configuration of the part being cast. Injection pressure, the presence and concentration of die release agent, and the degree of polishing of the die surface also influence shrinkage, but to a lesser degree.

Tables 6–8 define a different coefficient of thermal expansion for each of the commonly used die casting alloys. Dimensional or linear shrinkage in inches per inch, can be calculated when these values are multiplied by difference between the casting temperature and ambient room temperature. Then, if this number is multiplied by any dimension, the total amount of linear shrinkage to the length, width, and depth dimensions can be calculated.

Every casting cools from the outer surfaces inward, and there may be a considerable temperature differential between the outer skin, which is usually about 0.015 in. thick, and the internal mass of a die casting. The temperature gradient is less in thin sections than in more massive portions of the casting. The calculated shrinkage, therefore, is thus likely to be greater than the actual will be. For example, the calculated theoretical shrinkage for zinc-base alloys is 0.0096 in. per inch, but the practical allowance provided by most die casters and tool makers is 0.007–0.008 in. per inch.

All of these data are designed into the dimensions of steel die cavities since they operate at several hundred degrees and are manufactured at room temperature. The great difference between these temperatures must be allowed for calculating allowances for thermal shrinkage.

The amount of expansion (reverse of shrinkage) that the steel die material can be expected to experience in heating from room temperature to operating temperature must also be considered by subtracting this value from the shrinkage calculation. This calculation has been considered in the previously suggested practical shrinkage allowances.

The configuration of the part to be cast also has an effect. For instance, the 12 in. long unrestricted dimension, depicted in Fig. 11, will experience shrinkage much closer to theoretical than if the same dimension were confined at several points along its length. Estimating the extent of the effect of die restrictions is somewhat a function of experience, but one approach is to calculate the shrinkage with the limit of the longest free-standing dimension in

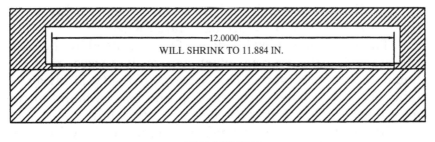

UNRESTRICTED

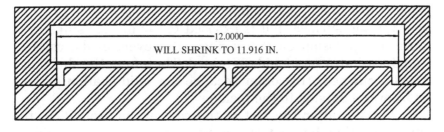

RESTRICTED

Figure 11

mind. The perspicacity and instinct of the tool makers have improved to an impressive level.

The zinc-base alloys experience phase changes that affect the volume by 0.0007 in. per inch. Of course, this factor is size sensitive and this is critical only for large dimensions.

All of these conditions should be considered for a complex casting, particular if it is large in size. The general shrink factors for the average casting used by most die casters and tool makers are:

Aluminum and magnesium alloys: 0.006–0.007 in. per inch

Zinc-base alloys: 0.007–0.009 in. per inch, depending on the alloy

Brass: 0.008–0.010 in. per inch

Shrinkage porosity affects the internal integrity of the casting when a void is created. This defect can be distinguished from gas porosity, the other cause for porosity, by examining the appearance of the pore. Shrinkage porosity always displays a rough and irregular inside surface since there is a dendritic structure associated with it.

The rough irregular inside surface is caused by the casting alloy literally tearing apart just as it passes from the liquid to the solid state. This type of porosity is always a function of the shape of the casting and can be found in the most dense region that is the last place to solidify.

Remember, this serious defect is generated by volumetric shrinkage while the semisolid casting is still contained between the steel die cavities. When shrinkage occurs in an open ingot mold, the cooling process is slow and gradual, and the top surface that is exposed to air merely sinks in to form a depression on the surface. However, in die casting, solidification is rapid and contained. Therefore, the shrinkage must occur near the center because the solidification pattern is from the outside to the inside.

Finally, after the rest of the casting has solidified, the only area that has not is the most massive one, possibly a heavy boss or a thick wall section that tears apart at the center because the surface has already solidified.

Thus, the internal integrity has been compromised by a void in the structure and the casting will leak or possibly fracture at this point.

Melting loss is caused by the affinity of most die casting alloys for oxygen. Whenever the liquid alloy is exposed to the atmosphere, an oxide skin is formed that must eventually be removed or skimmed away from the ladling site. This residue is called *dross* and signals trouble for casting quality, but is usually very rich in the base metal.

Even though there are various methods of rendering or extracting the base metal for remelting, a melting loss of between 4% per pass for a well controlled aluminum operation to as high as 20% for a poorly controlled magnesium melting program can be expected. Each time metal is remelted is called a pass.

Therefore, an alert manager will keep a watchful eye on the design of metal feed systems (runners) for an efficient balance between runner and casting volumes because all runners must be remelted, even in a perfect world. There is no rule of thumb for this audit, but the ratio is very sensitive to casting size. Larger castings usually have a smaller runner-to-casting volume ratio. As in everything else, the key is to get the best return of salable castings for the smallest investment in runners.

Cavitation causes small pits in the die surface near the gate in die casting from zinc alloys. These appear very soon after starting to take shots on a zinc die. The pits are approximately 0.010–0.015 in. in diameter and located about an inch downstream from the gate orifice where the metal exits the runner and enters the casting. The blemishes appear as small bumps as they are raised on the casting.

This phenomenon escalates with liquid density of the casting alloy. Thus, the heavier alloys of zinc, as opposed to the lighter alloys of magnesium and aluminum, are most affected. An analogy to a large semitruck making a turn at high speed to a small automobile in the same situation may be helpful to understand cavitation. Of course, it is more difficult to turn the truck since it is more massive and has a greater tendency to continue in a straight line than the car.

Like in the truck, the speed can be slowed down, which sometimes works, but at the expense of gate speed and cavity fill
time.

The generation of cavities in a fluid occurs when local
pressure falls below the vapor pressure of the fluid whenever
bubble nuclei are present. A bubble carried along in a stream
of liquid metal is not stable since local velocity and pressure
are continually changing (Karni, 1991). Bubbles normally collapse after a short lifetime. Often they collapse near the die
surface as depicted in Fig. 12. This is called an implosion
and frequent repetition at the same spot can cause serious
die pitting. The source of the bubble can usually be located
where the flow is more turbulent like a sharp bend in the
runner.

Many times die casters are surprised by this die pitting
when it occurs in zinc die castings because this material is
considered more gentle to die steel surfaces. The explanation
is that zinc is heavier and therefore resists any change in

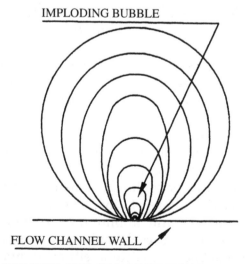

THE COLLAPSE OF A CAVITATION BUBBLE
TOUCHING THE WALL OF A FLOW CHANNEL

Figure 12

direction more than aluminum, which is lighter but much harder on the die steel.

Temperature drop is one of the dynamic events that occur during the die casting process. For purposes of this discussion, we will start at the superheat of the casting alloy as it leaves the breakdown (melting) furnace. The next stage is the holding furnace or pot where the temperature drops to about 40°F above the desired temperature at the gate for most cold-chamber operations. This drop is only a few degrees Fahrenheit above the gate temperature for most hot-chamber operations.

The difference is due to the method used to transport the alloy to the die. In cold chamber, the metal is usually ladled where it is totally exposed to the ambient room temperature; this causes it to drop dramatically in just a few seconds. In the hot-chamber process, the liquid alloy is transported through an enclosed hydraulic system where its only exposure to ambient temperature is on the surface of the bath.

Since die casting is really a thermal process where control of metal temperature is critical, this certainly puts the hot-chamber process in a favorable light. This gap between the two basic processes prompted the American Die Casting Institute, one of the forerunner organizations to NADCA, to participate in research to cast aluminum alloys by the hot-chamber process during the 1970s. The work cost over a million dollars before it was given up. The problem was not in the hot chamber of aluminum, which produced superb quality at very fast production rates. The proper composition for the plunger tip and gooseneck could not be found since most other materials dissolve so quickly in aluminum.

As the alloys travel though the runner system, which is no more than a conduit, the temperature continues to drop until the leading edge of the metal stream reaches the gate. At this location, velocity dramatically increases and the friction causes the temperature to increase. Then, during cavity fill, the constant drop continues until the cavity is completely filled. The total temperature drop in aluminum alloys exceeds 100°F at the end of cavity fill.

Finally, another rapid temperature drop occurs during the dwell phase of the die casting cycle during rapid solidification. The temperature of the casting that has been formed drops another 500°F before it has gained sufficient solid strength for ejection!

At the risk of redundancy, please remember that this rapid solidification is the one single thing that makes die castings unique from castings produced by other foundry procedures. It creates the fine, dense grain structure that die casting buyers seek.

Thermal constants define the behavior of not only the casting alloy, but the die steels that form the die cast shape and the cooling medium used for internal temperature control and external die spray. Computerized programs are available, in addition to consultancies, to make effective, predictable thermal calculations.

This is not simple except for the mathematical formulae and should not be attempted without a thorough understanding of the die casting process. Major improvement can be expected in casting quality and increased productivity can be expected if this job is done properly.

Too many of the die castings produced, even today, have not been subjected to this logic. How then are so many acceptable castings made, if this is so important? Well, the high pressure die casting process is so forgiving that many rules can be broken and castings can still be made that the end user can use. The predictability is considerably compromised, however. The yield rate varies from 70% to 95%, which is not acceptable for survival of the domestic industry given severe off shore competition.

This writer has been involved with many different dies, casting the whole array of alloys described here at many different die casting firms in North America, Europe, and the Asian basin and found only one die casting producer that experienced enough shots per hour. Most are 50% short of possible productivity. The single die caster broke all of the accepted thermal rules though.

The *liquid density* determines how the alloy travels through the system and its ability to change direction. The *liquid specific heat* is the highest temperature in the flat area of the metal cooling patterns illustrated earlier in this chapter. The *solid specific heat* is the lowest temperature in this flat or slightly sloped portion of the chart. The heat given up between the two is the *latent heat of fusion*.

It is the latent heat of fusion that determines how fast each alloy must be cast (cavity fill time). Using this criterion, Table 9 shows, that of the three major alloy groups, magnesium must be cast faster than the other two.

The liquidus is the temperature above which the metal is liquid, and the solidus is the temperature below which the metal is solid. Thus, the casting alloy behaves much like an hydraulic fluid above the liquidus and becomes slushy as the temperature drops. All analytical procedures for die casting are based upon the assumption that the temperature is above the liquidus, but it is necessary to constantly calculate the temperature drop to be certain of this. Premature solidification before cavity fill can cause cold shut, porosity, lamination, and poor fill without analytical strategy and control.

The injection temperature is the temperature of the metal as it reaches the gate, and is the only one of the thermal constants that can be changed in the die casting plant. All of the others have been designed into the alloy by the smelting operation and are therefore axiomatic as far as the die casting process is concerned.

The envelope of available alloys is stretching slowly and gradually. As this is written, traditional aluminum sand cast

Table 9 Thermal Constants of Major Alloy Groups

	Alloy group			
	Aluminum	Magnesium	Zinc	Units
Liquid Density	156.07	113.00	382.00	lb/cu. ft
Liquid Specific Heat	0.26	0.25	0.10	Btu/lb/°F
Latent Heat of Fusion	168.96	160.36	43.00	Btu/lb
Liquidus Temperature	1094.00	1103.00	726.80	°F
Solidus Temperature	1076.00	878.00	716.00	°F
Desired Injection Temperature	1200.00	1180.00	800.00	°F

alloy 356 is being die cast and heat treated to T4. Aluminum alloys are also die cast in thixotropic form as well as a semi-solid with the possibility of zero porosity because of the lower level of turbulence during cavity fill and reduced freezing range.

5

Metal Handling

The die casting process is initiated and depends upon a proper and adequate supply of casting alloy that meets specific requirements. The metal must have the correct chemical composition and be physically clean. Metal handling is the procedure of converting casting alloys from solid to liquid and back again in addition to moving metal between melting and casting stations. Furnaces are used to melt and hold the metal when it is in the liquid state. Therefore, much of this chapter covers the many facets of furnace design, construction, and operation.

Casting alloys are usually delivered to the die casting plant in the solid state. The raw metal can be in the form of small ingots that can be manually handled or large sows that must be mechanically charged into the breakdown furnace. Sometimes where production schedules allow, aluminum alloys are delivered hot in the liquid state. Of course, metal must be supplied, in the liquid state, to the die at a specific and acceptable temperature. The flow chart illustrated in Fig. 1 depicts the circuitous routing that must be carefully managed.

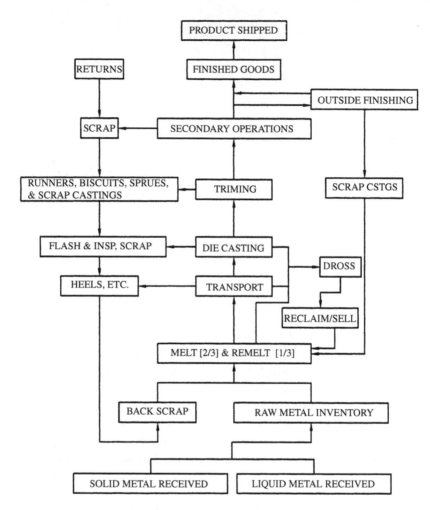

METAL HANDLING FLOW DIAGRAM

Figure 1

Metal has a different value at each station, and it behooves the alert die caster to audit them individually rather than merely assuming a percentage of metal cost to cover melting cost and loss for purposes of the cost estimate and financial statement. This is called activity based costing. Metal and heat energy are two of the three highest cost elements— the other being labor—in the manufacture of die castings.

The metal generally used for high pressure die casting is referred to as low temperature when compared to high temperature metals such as iron, copper, and silicon. The liquidus temperature of zinc alloys is approximately 700°F; aluminum, 1100°F; copper, 1900°F; iron, 2200°F; and silicon, 2600°F.

To meet these requirements, adequate melting and holding equipment is important. Superheated liquid metal must be scheduled to arrive at the holding furnace at the casting machine on a specific schedule determined by the volume of metal being processed.

Die casting is a thermal process and the superheated casting alloy is the heat source for the casting process. Natural gas or electricity provides the energy to superheat the metal. The chart in Fig. 2 describes the heat content of typical casting alloys at different temperature levels. For perspective, there is considerable thermal difference between the zinc alloys and higher temperature aluminum and magnesium metals.

The control of this heat energy is the key to productivity and quality. This chapter will discuss liquid metal containers, liquid metal treatments, heat sources, and thermal controls. The purpose of each container is to hold the charge while it is being melted, hold the melt at a designated temperature, or to provide a means to transfer the melt. Furnaces, ladles, launders, troughs, and crucibles all contain casting alloys in the liquid state while performing their other functions. They are designed to withstand the erosive action of superheated metal upon the material of the container.

Approximately 1000 btu of energy is required to melt a pound of aluminum and it can readily be observed from the chart that only a small portion is absorbed by the metal. This huge loss of heat energy is expensive and extremely destructive to materials that come into direct contact with it. Most of the wasted energy goes up the stack in the form of hot 2500°F air. Modern melting furnaces are designed to use some of this hot air to preheat ingots or sows before moving them to the melting chamber.

There are many sources of heat available, but the most prominent in die casting are natural gas and electricity,

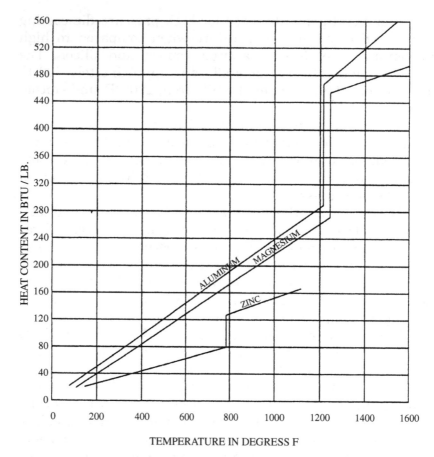

TEMPERATURE IN DEGRESS F

HEAT CONTENT OF TYPICAL CASTING ALLOYS

Figure 2

which will be the focus of this discourse. The utilization of this
heat energy is very inefficient since only a small portion is
actually absorbed into the casting alloy, as illustrated above
in the heat content chart for different casting alloys at different temperatures. This is a serious concern because energy is
the third highest cost of producing die cast components, after
the cost of metal and labor.

The die casting process can be viewed as a method to
exchange heat. This exchange starts with superheat generation to convert metal from the solid to the liquid state. Once

liquid, the metal has absorbed some of the heat energy and the exchange is completed when the metal is again converted, in the form of a net shape, to the solid state.

This heat exchange is extremely inefficient and can be stated by the formula $E = H1/H2$ where E is efficiency, $H1$ is heat in, and $H2$ is heat generated (Mangalick, 1976). The efficiency is usually about 50%, which means that half of the expensive heat energy is wasted.

ABOUT FURNACES

The type or style of furnace usually depends upon the alloy to be melted or held in the liquid state. There are also other points to consider such as energy efficiency, metal quality, capital investment, and operating cost. As with everything else in die casting, there are trade offs where negative conditions must be accepted to get to the positives.

The function of the furnace is the basic conversion of the metal from the solid state to the liquid state so that it can be used to produce die castings. Then, the die casting process returns the metal back to the solid state after it is converted into a usable shape.

Design characteristics of furnaces include:

• Reasonable construction cost
• Competitive operating energy costs
• Provision for efficient interior cleaning
• Adequate capacity to supply casting machines
• Acceptable service life of the refractory liner

Furnaces are designed for two purposes. First, there is the melting function. Melting is also referred to as the break down. After melting, liquid metal must be inventoried until it is transferred to the die casting die. Furnaces utilized for this purpose are referred to as holding furnaces.

Central break down furnaces for each alloy are recommended over melting metal and holding it at the casting machine. This is the only method that will maintain desired metal temperatures of liquid aluminum at the gate.

The metal is cleaner, but that issue is secondary to thermal control.

Holding furnaces receive metal from the break down furnace and are located adjacent to the shot end of each casting machine. From there it is ladled into the pour hole of the cold chamber. It is not a good idea to charge room temperature ingots, which defeats the tight thermal control.

Sometimes die casting firms choose to break down their metal in the holding furnace, which eliminates the need for central melting. This is a dangerous strategy because fluctuations in temperature are too drastic and too frequent. In such a scenario, preheating ingots are essential and many of these furnaces are designed with a melting hearth to reduce the range of temperature deviation from the ideal.

It is axiomatic that metal in the liquid state be available at all times. This requirement almost makes it mandatory that a die casting operation be continuous on a three-shift, 24-hour per day basis. Whenever metal is maintained in the liquid state without supplying production, energy is wasted and metal is lost through oxidation. It is customary, however, to keep aluminum liquid over weekends when no production is going on because it is too costly to remelt or break down after such a short shut down. Zinc can be allowed to freeze solid over the weekend and magnesium is vulnerable either way because of its propensity to oxidize.

The reverberatory furnace type (Jorstad, 1985) is often the choice for die casting aluminum. This furnace type is more robust and less sophisticated than others. Therefore, seriously deteriorated conditions can be tolerated. This, in no way, should be taken that the reverberatory furnace is the best choice. It suffers from low energy efficiency which is normally in the range of 20–25% at best. Heat losses from the flue gases and products of combustion are considerable.

The reverberatory furnace is basically a container, not only for liquid metal but for heat energy. It has to prevent as much heat from escaping as possible and must facilitate heat flow into the melt. Thus, heat losses need to be minimized.

In the reverberatory furnace, flames of combustion (usually gas fired) transfer heat to the metal by radiation and convection. The basic chemical conversion is the combination of carbon and hydrogen from the fuel with oxygen from the intake air to form carbon dioxide and water vapor.

There is no danger from the water thus formed since the temperature is high (>2000°F) and the pressure is low. Thus, the water vapor does not condense before it exits through the flue.

The flames are directed across and at least 12 in. above the surface of the metal bath in horizontal paths. Sometimes, however, holding furnaces can be seen with burners in the ceiling that impinge directly upon the surface of the bath— this generates far too many oxides that end up as dross and increases the melting loss.

Normally, this type of furnace is rectangular in shape—a box, if you will—but it can be any other shape such as circular, like rotary furnaces. In the rotary style, a large refractory-lined horizontal cylinder is merely rotated to pour off the liquid metal. Figures 3 and 4 illustrate some of the features described here for the two styles.

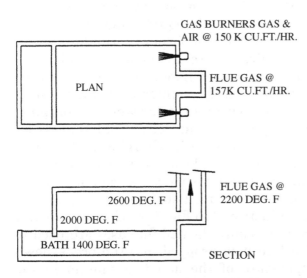

SCEMATIC DESCRIPTION OF REVERBERATORY
MELTING OR HOLDING FURNACE FOR ALUMINUM

Figure 3

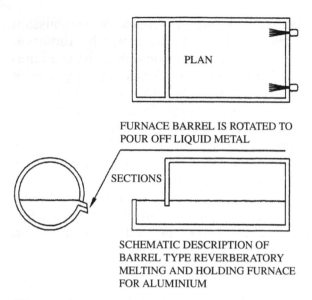

SCHEMATIC DESCRIPTION OF
BARREL TYPE REVERBERATORY
MELTING AND HOLDING FURNACE
FOR ALUMINIUM

Figure 4

The barrel shape makes pouring more convenient. However, the refractory wall, which is exposed to superheated air when stationary, is washed with liquid aluminum during each pour. Aluminum oxide in the form of corundum builds up rapidly, which requires more frequent removal. It is an ugly job that must be done by hand, so there is a strong tendancy to procrastinate orderly maintenance of the walls.

This type is provided with an outside well for charging, which may be located anywhere on an outside wall that is convenient for the die caster. The exterior charging well reduces metal loss by exposing only a small surface area of the bath to the atmosphere. Fluxing and drossing can also be performed without disturbing the interior melt surface.

The stack melter illustrated in Fig. 5 in schematic clearly describes how some of the superheated flu gases are used to preheat the metal before it is dropped into the break down chamber. The temperature of the liquid aluminum in the holding bath is 1300°F. Just prior to that, the temperature ranges from 850 to 1000°F. A metal temperature of 650°F reached on the preheating grill is over 500°F above ambient

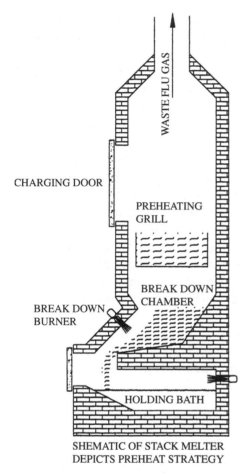

SHEMATIC OF STACK MELTER
DEPICTS PREHEAT STRATEGY

Figure 5

room temperature, which explains the high improvement in melting efficiency. Still the waste flu gas escapes to the atmosphere at almost 500°F, so the performance is far from perfect.

This furnace design claims to be 52% efficient with 48% of the energy wasted. Even so, compared to 25% efficiency reached with reverberatory furnaces, the performance is remarkable.

Crucible type furnaces are used to melt and hold zinc, and sometimes magnesium casting alloys. The crucible or pot is usually cast iron. This type should not be used with aluminum alloys because of their solubility for iron—an iron crucible will be dissolved by the liquid aluminum it holds.

The hot chamber holding furnace is an example of this type where the cast iron gooseneck is immersed in the bath of liquid zinc or magnesium. A schematic of this type of furnace, in which the flame and hot gasses from the burner circulate around the chamber between the refractory wall and the crucible, is offered in Fig. 6.

Immersion tube burners are used extensively for melting zinc. Figure 7 describes how the gas flame circulates within the tube that is immersed in the bath of liquid zinc.

When magnesium is the alloy to be cast, it is necessary to add a cover to the crucible and also a cover gas, usually SO_4.

Electric induction furnaces are used to hold liquid aluminum where the advantage is the circulation of the alloy to better keep all of the alloying elements in suspension. A magnetic field is created within the liquid metal bath directly.

Eddy currents in the melt generate heat and directional forces which in turn cause the desirable circulation. The disadvantage with this type of furnace is that it is too fragile for most die casting operations.

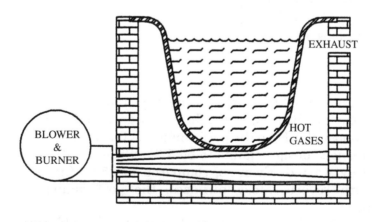

CRUCIBLE TYPE FURNACE

Figure 6

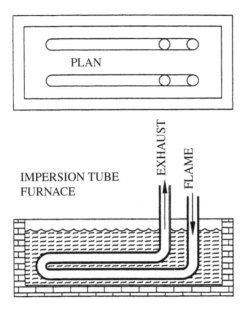

PLAN

IMPERSION TUBE
FURNACE

EXHAUST

FLAME

Figure 7

The energy efficient glow bar furnace uses electric power as the heat source; the furnace container is made from highly insulative nonmetallic board. Electric glow bars located in the roof radiate heat onto the surface of the bath. This type is used only for holding liquid metal at the casting machine.

Even though the energy savings and reduction in melting losses are significant with this type of furnace, constant and continuous maintenance is required to keep buildup of oxides from forming. This tight preventive mainte-nance is beyond the ability of some die casters (B. Guthrie, 1995).

Adequate capacity can be established with finite accuracy if all necessary factors are considered (Table 1). With all fur-nace types, the melting or holding capacity is determined by engineering calculations that address:

- Hearth area
- Burner capacity
- Air requirement
- Metal depth

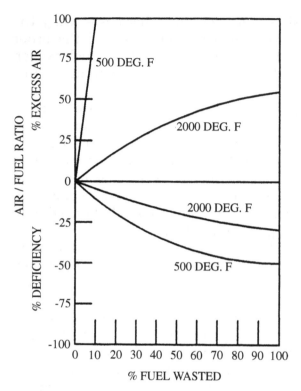

Figure 8

recuperation in which potential fuel savings are presented (Altenpohl, 1981).

Acceptable service life of the refractory liner is crucial to the performance of the reverberatory furnace type as well as all others. The choice of material and subsequent care and maintenance are important. This is not the place for cheaper options.

Refractories must withstand the physical load at the high operating temperatures. The material has to have a low thermal conductivity, and it cannot react chemically with the melt, dross, or flux. Alumina–silica materials are used, which are acidic. pH increases with alumina content, which is neutral at 60%. Silicon carbides have a neutral pH, and magnesium oxide materials are basic in nature.

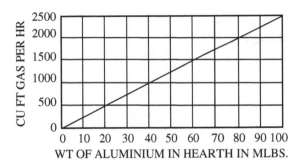

WT OF ALUMINIUM IN HEARTH IN MLBS.

Figure 9

A key factor that limits the chemical reaction is the acidity of the refractory material. The acidity must be matched to that of the metal oxide of the casting alloy, and the flux that will be melted and contained in the furnace.

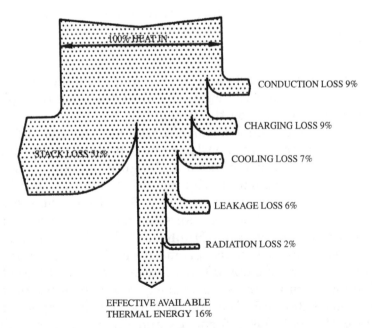

HEAT BALANCE IN CONVENTIONAL FLAME-HEATED FURNACE

Figure 10

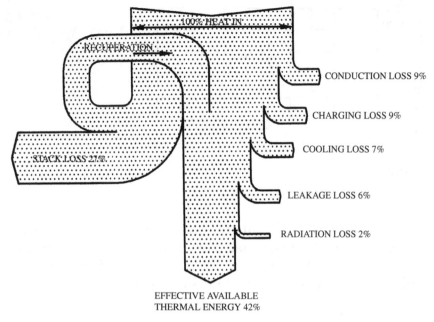

EFFECTIVE AVAILABLE
THERMAL ENERGY 42%

HEAT BALANCE IN FLAME-HEATED FURNACE WITH RECUPERATION

Figure 11

All refractory bricks and mortar must be of nearly equal acidity. If this is not the case, chemical reactions take place that will quickly destroy the furnace lining.

Maintenance and interior cleaning are onerous and hostile tasks that are easy to postpone but absolutely essential. If a well-planned preventive maintenance schedule is not enforced by management, the condition of metal handling equipment deteriorates very rapidly. Sometimes the maintenance schedule is allowed to lag so as not to adversely affect short-term operating profit. Haphazard cleaning and maintenance wears out a furnace prematurely. Then, if quarterly earnings of say $500,000.00 are forecast, and a $100,000.00 maintenance program is postponed, reported earnings can be enhanced 20% or 5% if amortized over a year! Proper maintenance of break down furnaces and other metal handling equipment can be held off almost indefinitely, but energy cost and metal quality suffer as a result. Some managers do not think of it this way, because they merely believe this expense is not affordable.

Consequently, one does not have to look too far into the die casting industry to observe worn out furnaces that spue out open flames from cracks and poorly fitting doors. Ignoring or postponing cleaning to cut costs or because it is difficult also creates a hostile environment in the rest of the foundry. The die casting and remelt departments end up looking like Dante's inferno ... a hell of a place to work. Conversely, properly maintained melting equipment establishes a neat and efficient operating climate that enhances productivity and safety.

The thermal conductivity of the refractory is critical to this task. A K factor quantifies the conductivity each refractory in terms of btu/hr/sq.ft/°F/in. of depth. A high K factor specifies a high conductivity. Since the function is to contain heat, a low factor is desired. Unfortunately, though, the best materials for containing liquid metal have high thermal conductivity.

Therefore, to minimize heat loss, refractories are backed up with highly porous insulating materials. Heat is always transferred from a high temperature source (burners) to a low temperature receiver (metal). In both melting and holding furnaces, heat is transferred by:

- Radiation that first transfers heat from the burner to the metal.
- Convection that transfers heat when fluid particles flow between hot and cold metal crystals.
- Conduction that occurs when heat flows from the hot face through the refractory and insulation to the cold side.
- Radiation that carries the heat away from the cold face.

The aluminum break down furnace presents a worst case scenario. It is absolutely necessary to scrape aluminum oxide from the refractory walls and bottom before it turns to extremely hard corundum. There must be no more than two days between routine cleanings. An especially vulnerable location is the metal line; Build up around the walls occurs as it is constantly exposed to aluminum in the liquid state, and

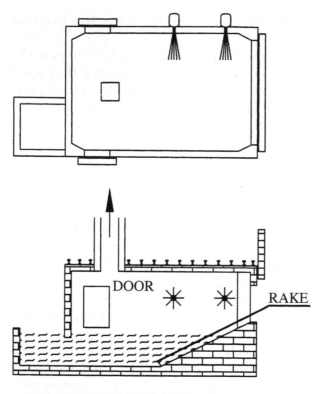

CLEANOUT RAMPS AND DOORS

Figure 12

oxygen and hydro carbons. This maintenance work can be considerably eased by simple features in furnace or holder design that have no adverse effect upon efficiency or productivity. The initial cost is higher, however, because the angle takes up bath space. The footprint of the furnace must then be larger. Figure 12 sketches angled clean out ramps at the bottom and corners. Note the strategically located doors for convenient chipping and raking of loosened dross.

METAL HANDLING

Metal handling should be minimized because it is only a necessary means to an end. It is costly to move solid metal from one place to another so some die casters receive it in liquid form, which is called hot metal. Costs are reduced since it has to be liquid for the smelter to alloy, so eliminating a solidifying and remelting cycle saves significant energy. Then, the best means of transfer should be used to get the metal into the die so that net shapes can be formed.

Good metal handling can minimize casting defects that are caused by impurities introduced during transporting of liquid metal. Nonmetallic inclusions contain oxides and spinels, which are complex double metal formations. Furnace refractory debris and sludge also contribute.

Corundum is a very dense form of aluminum oxide and eventually becomes hard spots in castings that can damage machining tools. Corundum conversion occurs under conditions of high temperature and inefficient combustion. Furnace refractories with higher silica and alkali oxides, especially sodium, contribute to the formation of corundum in the alloy.

Prevention of corundum is a function of wise choice of furnace refractory, good operating practice, and preventive furnace maintenance to remove oxide accretions while they are managable (Jorstad).

Aluminum alloys are often degassed and oxides are filtered out during the smelting operation, but the benefits are often lost to subsequent poor remelting and handling within the die casting operation. Metallic and organic residues in addition to moisture can be present in back scrap material.

Hydrogen gas can be absorbed by aluminum through incomplete combustion in fossil fuel-fired furnaces. Hydrocarbon residues from metalworking lubricants and hygroscopic fluxes can be present. Usually though, hydrogen pick up comes from the atmosphere and temperature of the liquid aluminum. The alloys are especially vulnerable to this phenomenon at temperatures in excess of 1400°F. It can also occur at lower temperatures during periods of high humidity.

Several methods are used to detect the presence of hydrogen in liquid aluminum alloys. Density, reduced pressure testing, vacuum fusion, and "hydrogen probes" are available laboratory techniques.

The high pressure die casting process is forgiving of some amount of gas content and most casting defects can be traced to causes other than metallurgical. However, the more enlightened die casting plants require clean metal, so a brief discussion on degassing methods is included here.

Historically, hexachlorethane-based tablets are submerged into the bath of liquid aluminum. This gas reacts with aluminum to form aluminum chloride which acts as a sparging gas to collect hydrogen. This method works well for small melts but the tablets are difficult to store, create obnoxious fumes, and must be plunged deep.

A better technique utilizes a lance to inject inert gases such as argon or nitrogen, or reactive gases like freon or chlorine into the melt. A fluxing tube is used but is inefficient since it produces large sparging gas bubbles. Large bubbles tend to coalesce quickly and thus do not disperse completely though the bath of liquid aluminum.

Porous ceramic plugs to dispense the sparging gas can be constructed into the bottom of the furnace or holder that work well. The problem is that they are too fragile and become clogged too easily. Of course this requires major maintenance and is not practical.

The favored method of degassing is the rotating injection system. A rotor is immersed into the melt and mixes the sparging gas with the aluminum while shearing the gas bubbles. The gas is uniformly dispersed and treatment times are significantly reduced. Figure 13 compares the efficiency of some of these degassing methods and clearly identifies the rotor or impeller, as it is usually called, as superior.

DROSS AND MELTING LOSS

Since metal is the highest single cost element in the production of die castings, it is important to get as much of it through the process and shipped out as salable product as possible. Dross naturally forms when casting alloys are melted. It is the physical evidence of melting loss. A loss of 3% in melting aluminum ingot is considered normal, but this can increase to

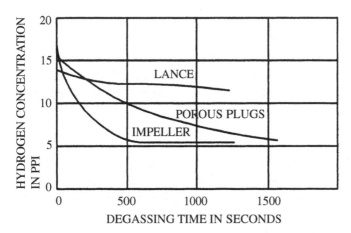

RELATIVE EFFICIENCY OF DEGASSING METHODS

Figure 13

as much as 15% when back scrap is remelted. The loss may be slightly lower for zinc and considerably higher for magnesium. These losses have a huge economic impact and are the result of too much dross generation during melting and holding.

True aluminum dross is really aluminum oxide (Al_2O_3) and usually consists of about 80% aluminum and 20% oxide. Even the untrained eye can detect dross that is too rich in metallic content because it shines more than low grade dross, which looks more dusty and gray in color. A large portion of the aluminum can be recovered by freeing it from the oxide and returning it to the liquid aluminum bath. This is accomplished by treating the dross with exothermic flux.

Fluxing to minimize dross that forms on the surface of the melt is an essential task in metal handling to prevent excessive melting loss. Flux materials used in die casting are mixtures of inorganic and sometimes organic salts.

Flux is a combination of chlorides, fluorides, and oxidizers that superheats the dross floating on the top of the melt surface. When applied at a rate of about 2 lb./sq. yd. of surface, this melts the aluminum content back into the base metal. The temperature during this process, while very high, is too low to melt the oxides, so they remain on the surface and can then be dragged off.

Fluxing salts are rabbled into the dross layer to release entrained metal back into the liquid bath. Metal droplets are stripped from their thin oxide skin and coalesce back into the main bath. The flux superheats the dross on the surface of the melt so that the aluminum is freed when it becomes liquid again. This process can be enhanced by mechanical agitation, sometimes through a heavy screen, or even when the dross is spread out in a refractory pan. The beneficial results, in addition to reduced melting loss, are better metal quality and cleanliness, improved product quality and machinability, and prolonged furnace life.

Sometimes flux serves only as a physical barrier as in the case of cover fluxes. In this case, flux is spread across the surface of a bath of liquid casting alloy to reduce hydrogen pick up from the atmosphere. This flux is usually composed of active fluoride salts. They are also used to absorb lubricants, dirt, and other debris in the charged metal, especially back scrap and trimmings.

Other fluxes may be mixed into the main bath to purge oxides and other impurities. The wetting that occurs causes agglomeration in which the impurities return to the dross layer on the surface. The difference in specific gravity defines a buoyancy in the agglomeration so that it literally floats up to the top, leaving clean liquid metal in the melt beneath.

Successful fluxing requires correct composition, proper quantity and application, and sufficient contact time. Flux and flux residue must be completely removed by thorough sedimentation, floatation, and skimming.

Elements that form dross can also be removed with an inert or active gas flux. Some foreign materials with lower densities than the casting alloy float to the top of the liquid metal bath and this is called dross. Others of higher specific gravity sink to the bottom to create sludge. Also, some undissolved foreign particles and/or gas may be suspended in the melt. However, the removal of undesirable suspended particles with gas is not efficient since large quantities are required. Fluxes are used to cover the melt surface to reduce oxidation losses and to separate the dross from the base alloy.

In die casting zinc alloys, the dominant mechanism is the intermetallic compound $FeAl_3$. Like aluminum alloys, zinc dissolves iron, but at a slower rate in the presence of the aluminum, which is alloyed as a hardener. The dross, being lighter than the zinc casting alloy, floats to the surface of the melt where it is skimmed off.

Conventional melting of magnesium, the other major die casting alloy, is carried out in steel crucibles since there is no attraction for iron as with the others. During and after melting, a salt flux is used on the surface of the melt to prevent burning and to coagulate oxides that settle to the bottom.

Filtration, sedimentation, or floatation removes solid particulate from liquid aluminum (Neff, 1991). Sedimentation allows heavier impurities to sink to the bottom of the melt where they may be dragged out or restrained from entering the pouring stream. Foreign materials that are lighter than the aluminum will escape by the mechanics of floatation during degassing.

Many aluminum oxides do not differ significantly in specific gravity from the aluminum matrix. In this case, the only way to keep them out of the pouring stream is mechanical filtration.

Nonmetallic inclusions can be filtered out with a fiber glass or porous ceramic device strategically located in the pouring stream. Filtration is accomplished by one of two mechanisms in which the impurities either become trapped on the inlet side or within the body of the filter. Since the filter will eventually become clogged, it is essential that it be replaced on a precise schedule.

The probability of removing all nonmetallic inclusions is not absolute because the assumption that their size is greater than the filter openings is not a certainty. The filter in Fig. 14 prevents nonmetallic impurities from moving from the break down chamber to the ladling well, which significantly improves the cleanliness of the liquid aluminum. Particles may just build up on the inside of the filter or become trapped within it, depending on the type. It is easy to see that metal

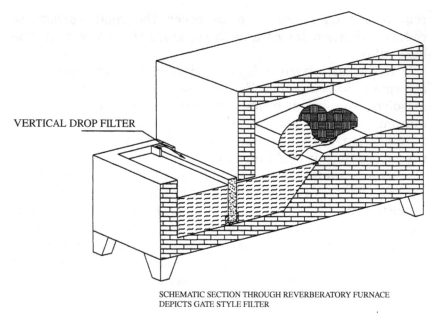

VERTICAL DROP FILTER

SCHEMATIC SECTION THROUGH REVERBERATORY FURNACE
DEPICTS GATE STYLE FILTER

Figure 14

will eventually cease to flow into the ladling well if the filter is
not changed on schedule.

Magnesium alloys call for more complex handling
because they have a strong tendancy to oxidize and are flam-
mable in some forms. Uneducated people fear superheated
magnesium in the liquid state, but the danger occurs in the
form of chips, shavings, or dust. It can burn intensely and
appears white hot. Water only exacerbates the fire; only sand
will extinguish the flame.

These alloys are more difficult to melt and handle when
liquid because this metal oxidizes so readily. This affinity for
oxygen requires measures that are somewhat awkward when
compared to handling aluminum and zinc.

Material such as SAE 1020 steel or 430 stainless steel is
used for melting and holding crucibles because they do not
contain nickel (Koch et al.). Only high density, high alumina
refractories are compatible with magnesium. Low density,
high silica materials should be avoided.

It is also important to preheat ingots thoroughly to about 350°F to remove any moisture that might condense. This can be accomplished from totally automated devices to simply stacking them on top of a furnace.

Rapid oxidation is prevented by a protective gas atmosphere such as sulfur hexafluoride in air. Care must be taken not to unnecessarily disturb the surface so ingots must be submerged gently. Each time the surface of the melt is broken a new protective skin forms that contributes to dross accumulation.

Reactive gases from the vapor space above the melt must be excluded and vaporization of magnesium has to be suppressed (Baker, 1989). Sulfur hexafluoride is usually used since it is innocuous and only mildly corrosive in concentrations used for liquid magnesium protection. It is used in combination with air and carbon dioxide, or just air. Recommended concentrations required to protect commonly used alloy AZ91D are noted in Table 2.

The overall gas flow rate is as critical as the concentration of SF_6 in the gas mixture. A rule of thumb is three times the volume of vapor space above the melt per hour. This gas is expensive and should be monitored carefully.

A typical gas distribution system incorporates an air dryer, pressure and flow regulators for each gas, a mixing chamber, an in-line gas analyzer, a distribution header to each furnace, and a flow meter at each furnace. Fluxes must be kept out of this system because they reduce the degree of protection and corrosion resistance of the magnesium alloy.

Magnesium back scrap is more complicated than the other more commonly die cast alloys. Its affinity for oxygen requires fluxing that is severely corrosive and that requires

Table 2 Recommended Sulfur Hexafloride Concentrations

Melt temperature (°F)	No surface agitation Volume % SF_6	With surface agitation Volume % SF_6
1220	0.02	0.04
1301	0.04	0.12
1400	0.05	1.0–1.5
1499	0.06	Poor at all concentrations

remelting in a separate facility. Selling to or employing the smelter to render it is an option. Another choice is to remelt it through the original break down furnace but this adds to the degree of difficulty of controlling contaminants like iron and oxides. There is, however, a flux-free process that almost eliminates the other hazards, which suggests the best method.

TRANSFER

There are many methods to transfer liquid casting alloy and each has advantages and disadvantages.

Hand ladling is the most basic and simple. It offers flexibility, which is desirable, but it is too labor intensive for transfer between holders. When aluminum is hand ladled into the cold chamber, it takes a skilled operator to reproduce the same quality shot after shot, which is very unpredictable.

Gravity metering requires no labor and can be readily automated. Since there is no mechanism required, there are no moving parts. Many disadvantages can be expected, though: Transfer tubes must be heated to keep the metal liquid; metal levels are extremely important; and, of course, the holding furnace must be elevated. You can bank on valves leaking. This method will not work for aluminum castings that weigh less than 1lb.

Gas displacement has all of the advantages of gravity and all of the disadvantages as well, except that the holder need not be elevated.

Centrifugal pumping can move large volumes of metal rapidly and it can be easily automated. No labor is necessary and metal levels are not important. Transfer tubes must be heated and moving parts must be maintained. If this method is used to deliver metal to the cold chamber, the shot volume must exceed 10 cu.in.

Siphoning of liquid zinc is very effective but does require labor to initiate the siphon and also to stop the procedure when the job is complete.

Automatic ladling has been proven in production and widely used for cold chamber aluminum casting because there is no size limitation on the shot and the direct labor is eliminated. Ladle wash is used to keep the metal from sticking to the ladle. This transfer method is much more risky for magnesium and the trend is to go to the hot chamber process for automation.

Bull ladles are the usual containers utilized to transport liquid metal. They are carried by an overhead hoist or a specially adapted lift truck. The casting alloy is tapped from the break down furnace into the ladle and then poured out into the holder at the casting machine. Care must be taken to minimize splashing of metal, which is difficult to completely avoid.

Launders require a high capital expenditure, but transfer liquid metal from the break down furnace to the casting machine almost without labor or incident. A launder is an insulated and heated covered trough between the melting and holding station. It must be lined with a compatible refractory and temperature must be rigidly controlled. Metal levels in the casting machine holders are ensured and splashing is eliminated.

Some disadvantages of launders are that they restrict access to casting machines and require more heat energy to operate. Of course, only one alloy can be carried at a time.

CHEMICAL COMPOSITION

There are many ways that raw material received in compliance to specification can become contaminated with foreign material or other metals. Therefore, some means of control are necessary. All smelters provide a spectrographic service to their customers but there is a time gap that could be problematic. Most die casting firms have equipment in their quality control laboratory so that chemical analysis can be performed in house.

SAFETY

Any discourse on metal handling in die casting would be remiss without noting the safety hazards involved. Liquid metal splashes are capable of inflicting life threatening injury. In addition, all tools can be too hot to touch and thus can cause burns.

It is important that metal handling equipment be maintained in acceptable operating condition as defined by manufacturer's specifications. Safety systems on energy supplies should be checked regularly and maintained in acceptable operating condition. The lining and structure of all melting equipment must be monitored constantly by personnel to whom this responsibility is specifically assigned. Proper burner adjustment is necessary to preclude erratic ignition and combustion.

Water on the surface of a liquid metal bath merely bubbles to form steam and then evaporates. However, given the opportunity to get below the surface, it converts to steam and the rapid expansion generates a violent explosion powerful enough to blow all of the metal out of the furnace! Even partially filled soda cans discarded in a scrap container area are a source for moisture. It is important to preheat all tools so that they are uncomfortable to the touch before submerging them into a metal bath.

Ingots or sows should be gently lowered into the bath and not dropped. When liquid metal is poured, splashing must be minimized. Splash guards should be provided wherever liquid metal is poured.

Personal protective gear includes full face shields, heavy sleeves, hot mill gloves, spats, and safety boots. A system of enforcement is necessary since safety will be discarded if left to human nature.

WHERE DOES SUPERHEAT GO?

Since the injection temperature of the casting alloy as it reaches the gate has such a profound effect upon casting quality, and because the loss of superheat is difficult to determine in the cold chamber process, the nomographs included in Figs. 15–17 and at the end of this chapter should be helpful when aluminum casting alloys are used. (The transfer of liquid metal in the hot chamber process is so rapid that the heat loss is not significant.)

It is important to understand that these nomographs present a very simplistic approach and that they are included here only as an assist, which is significantly better than merely guessing at the loss in superheat that occurs in the cold chamber process prior to injection.

AN EXAMPLE

The application of the nomographs is illustrated in the discussion that follows. Starting with a holding metal temperature of 1225°F, which is fairly typical for aluminum die casting alloys, Fig. 15 is based upon a ladle radius of 4 in.

A construction line is then drawn between the 1225°F temperature and the 4 in. radius and extended to the reference line.

Then, another line is drawn between the reference point and the ladle time of 7 sec. This line crosses the heat loss line at 5°F, which means that the metal has lost 5°F during ladling.

Therefore, the temperature is 1225 − 5 = 1220°F at this point.

The second nomograph, Fig. 16, is used to determine the temperature drop that occurs during the time that the liquid metal is in the shot sleeve. Note that the scales are based upon the shot sleeve being at least 35% full, which is a good

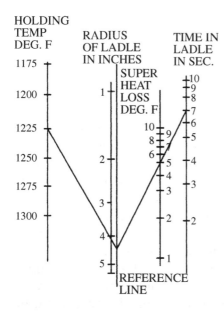

NOMOGRAPH FOR SUPERHEAT
LOSS IN LADLE - EXAMPLE

Figure 15

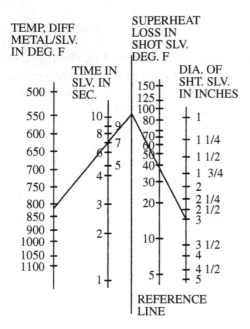

NOMOGRAPH FOR SUPERHEAT LOSS
IN SHOT SLEEVE (35 TO 75% FULL)

EXAMPLE OF APPLICATION

Figure 16

guide to follow in choosing the sleeve size to minimize air
entrapment at this time during the casting cycle.

Care must be taken to minimize wave formation during
the slow shot phase when the plunger tip starts to move. In
other words, the critical slow shot chart should be followed
so that too much aluminum skin does not form on the sides
of the shot sleeve in front of the tip. Of course, this skin is
scraped off and ends up as oxides within the casting.

This phase is especially important because the most
superheat is lost here. In the example, the difference between
the temperature of the shot sleeve material and the liquid
metal is estimated at 820°F. The sleeve temperature is therefore about 400°F, which can be measured with a simple
pyrometer.

A construction line is drawn between the 820°F point on the left scale and the 7 sec point on the second scale, which is the time that the metal is in the sleeve. This line is extended to the reference line. Then, another line is drawn between this point and the diameter of the shot sleeve, which in this case is 3 in. This line crosses the superheat loss line at 38°F.

Thus, the 1220° temperature after ladling losses can be expected to lose another 38° F, making the temperature in the cold chamber $1225 - 5 - 38 = 1182°F$.

Finally, heat is lost while the liquid metal travels through the runner as demonstrated in Fig. 17. Much of this

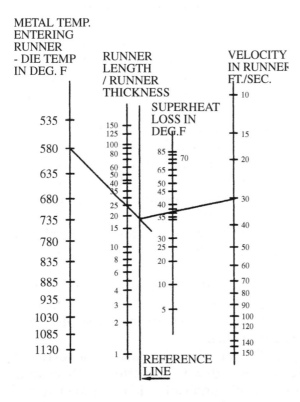

NOMOGRAPH FOR SUPERHEAT LOSS IN
THE RUNNER SYSTEM - EXAMPLE

Figure 17

event takes place during the slow shot phase of the casting cycle, so the velocity is relatively slow.

Again, the temperature gradient between the die surface and the casting alloy is a prime factor in calculating the freezing schedule as the metal approaches the gate. Therefore, the first scale is used to establish this ratio and, in the example, 600°F is estimated.

The second scale is designed as a quick way to estimate the surface-area-to-volume ratio. It loses accuracy since only the length and thickness of the runner are used rather than area and volume, but this is not serious since the loss of superheat in the runner is not too great.

The runner length is divided by the thickness and, in the example, the ratio is 25. A line is drawn through these two points to the reference line.

The velocity of the metal flow in the runner must be calculated either manually with the hydraulic formula $Q = AV$, where Q is quantity in the runner, A is the area of the runner, V is the runner velocity. When some CAD software is used, this calculation is automatic.

In this case, the velocity is 30 feet per second, and the line between this point on the fourth scale and the point established on the reference line crosses the heat loss line at 8°F. The metal temperature has already dropped to 1192° in the second nomograph; the temperature at the gate is calculated by the final formula of $1225 - 5 - 38 - 8 = 1174°F$.

Usually, a temperature at the gate closer to 1200°F is desired, so this scenario cannot be considered optimum and should thus be enhanced.

These same nomographs, without the work lines used for this example, are repeated here, in Figs. 18–20 so that they may be used by the reader as tools to determine the best holding temperature for specific dies.

An efficient layout for a typical die casting department integrates the melting (breakdown) operation with casting to provide a smooth flow of material. This is important if the modern management philosophies of just-in-time delivery, and minimum inventories are to be achieved. Since many bottle necks develop easily in the molten metal and casting

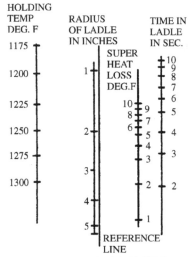

NOMOGRAPH FOR SUPERHEAT
LOSS IN LADLE

Figure 18

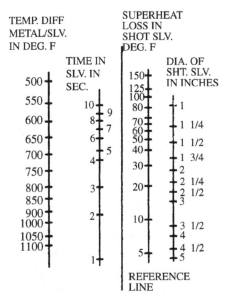

NOMOGRAPH FOR SUPERHEAT LOSS
IN SHOT SLEEVE (35 TO 75% FULL)

Figure 19

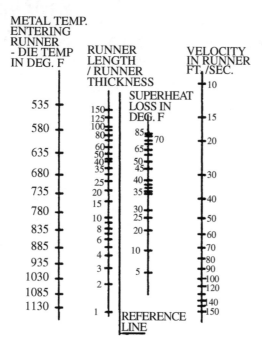

NOMOGRAPH FOR SUPERHEAT LOSS IN
THE RUNNER SYSTEM

Figure 20

systems, a simple and efficient layout for smooth material
flow is presented in Fig. 21.

When studying this layout, one should keep the first
chart in this chapter in mind because this puts some of the
routing into real time perspective (Fig. 1, p. 140). Note the
dross rendering station, which is the equipment to force the
80% of dross and skimmings back into usable casting alloy
and sell off the lean dross to the smelter.

Backscrap is usually delivered to remelt via an under-
ground conveyor that is designed with a minimum of obstructions for ease of maintenance. The course of the conveyor is
close to the trim presses since they are where the casting is
separated from the runners biscuits, sprues, and overflows.
The terminus is directed onto a charging apron so that it
can be easily charged into the furnace for remelting.

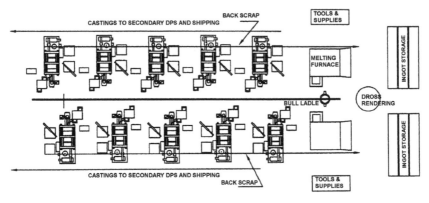

SIMPLE METAL HANDLING SCHEMATIC LAYOUT

Figure 21

Castings may be loaded into a tote for delivery to the secondary operations, or preferably onto an overhead conveyor for labor free transport.

Sometimes the bull ladle used to transport the liquid casting alloy from the melting furnaces to the holding furnaces at casting is handled with a specially designed lift truck, but since this method can be a bottle neck, the monorail is suggested.

6

Concepts of Cavity Fill

It is necessary to displace the air in the cavity with super heated liquid metal in order to convert the ingot of casting alloy into a useful shape. Most issues with cavity fill concern the metal, but it is just as important to deal with the air in the cavity that is at atmospheric pressure when the plunger starts to move. At the end of cavity fill, the air, if not exhausted from the cavity, becomes compressed. In this case, the compressed air pressure retards the flow of metal, especially at the end of cavity fill.

Vacuum systems are commercially available and work well if properly designed into the die. Natural air venting is more economical and almost as effective. The important thing is to size the vent area that sees the atmosphere, proportionally to the gate area. It should be noted that, under the high pressures used in die casting, a generous portion of air escapes from the cavity through the spaces between die components such as ejector pins and core movements, but most of all die blow between the two die halves.

Many studies have confirmed that the volume of air that must be displaced during every shot is much greater than merely the volume of the cavity. In all cold chamber dies that cast a volume of metal less than 400 cu. in., it is very difficult to maintain a 40% fill level in the chamber, which means that 60% of the total volume is air! This volume of air plus the air in the runner and in the cavity must be displaced by the metal before the casting shape can be formed.

The objective is to design the vents so that the air will exhaust from the cavity in front of the super heated metal stream at approximately the speed of sound in air. The speed of sound in 500°F air is 1608 ft/sec (19,284 in./sec).

A rule of thumb is to size the vent between 10% and 20% of the gate area. Most die casting dies do not provide enough geometry to achieve the latter, so 10% or less is usually what one sees in most dies.

Since low cavity fill time is desirable to maintain the casting alloy above the latent heat of fusion during fill, a larger gate area is one way to accomplish this without excessive gate speeds. Most die designers overlook the effect that this strategy has on the vent area. Therefore, too many die casting dies are under-vented.

Cavity fill is dependent upon the thermal behavior of the superheated casting alloy that is in the liquid state during this brief but critical period in the die casting cycle. It is important to understand that the physical state of the alloy changes from liquid to solid before most cavities are completely filled. Actually, the state is more plastic since the metal cooling pattern is in the latent phase of transformation at this time.

In response to this rapid freezing phenomenon, considerable attention must be paid to the time taken to fill the cavity in an effort to effectively manage the physical state of the metal. The thermal constants of the various casting alloys used in die casting (discussed in Chapter 4) have a profound effect upon the condition of the liquid metal during cavity fill. They define the thermal behavior as the super heat is lost mostly by conduction to the die steels.

When the length of time required to fill the cavity is known, then the temperature of the metal at the end of cavity fill can be calculated. This is done by comparing the time it takes for a certain alloy to cool from its liquid specific heat and to reach the solid specific heat. This event usually takes place during cavity fill and is established by the latent heat of fusion.

The temperature of the die steel at the cavity surface provides the thermal environment in which this whole scenario takes place. Therefore, it is critical to calculate the most hospitable temperature for the specific net shape to be die cast and maintain it within a reasonably close range.

The shape to be cast defines the cavity contours and dimensions. It is the key to how far the limitations of the die casting process need to be stretched. This factor is usually not within the control of the die caster, but is determined by the product designer and the end use requirements of the casting.

The ratio between the volume of the cavity and the surface area is determined by the design of the shape to be cast. A quick way to appraise this factor is to measure the wall thickness of the casting since a thicker wall will hold heat longer than a thinner section. Most product designers opt for a thin wall to minimize the cost of the casting alloy. This determines the degree of casting difficulty during cavity fill.

Lower ratios of volume-to-surface area are more challenging because issues with surface finish quality can be expected. The tool engineer must design gating, venting, and process parameters to avoid cold shut defects. This is especially true for hardware zinc castings with cosmetic requirements.

The distance that the super heated liquid casting alloy must travel after it exits the runner through the gate and arrives at the last place to fill has just as profound an effect upon casting quality. Experience by this writer in gating over 400 different dies suggests that with normal wall thicknesses, the maximum distance for aluminum alloys is approximately 8 in. Since the freezing range of zinc and magnesium is shorter, this critical distance is about 4 in. Success requires the shortest possible cavity fill times.

The pattern that the streams of liquid metal travel during cavity fill is just as important to casting quality as the thermodynamics. Think of most castings as shells and it will be easier to visualize the fill pattern. The streams of metal follow specific routes after they exit the runner through the gate and enter the cavity. There are definite reasons that these patterns are formed and the available paths must be carefully studied so that the routes can be strategically planned. Too often, however, it is developed in a random manner. Die casting is a high speed, high pressure, turbulent process, so it is assumed that the metal stream travels in a straight line until it collides with some obstruction, such as a core within the cavity. Thus, it is important for the die caster to incorporate the course of the metal streams into the design of the fill pattern by calculating runner areas and locating gates so that the optimum direction of flow is achieved.

Some basic assumptions influence the fill pattern. Since the cavity fill time (measured in milliseconds) is extremely short, it is assumed that the viscosity of the super heated casting alloy is approximately equivalent to that of water. This is the reason for so many early water analogy studies that taught us so much about fluid flow.

Super heated casting alloy can then can be expected to behave as a hydraulic fluid during cavity fill. It is therefore critical that the temperature is kept above the liquidus of the alloy being die cast during this period of the casting cycle. By the same token, it is also assumed that the streams of liquid metal travel in straight lines until they encounter an obstruction.

Two flow theories somewhat explain the flow directions. One theory expects a frontal flow directly from the gate inlet to the extremity of the cavity. This condition is illustrated in Fig. 1, but note that the metal streams back fill along each side. Arrow heads indicate the direction of flow on the vectors. A particular advantage of the tapered tangential runner design is a unique ability to foster a full frontal fill pattern.

The other theory says that the metal will splash off of a physical detail or the cavity extremity and then backfill the cavity with liquid metal. Such a pattern is portrayed in Fig. 2.

Actually, some of each theory occurs during the short period of time that the cavity is being filled.

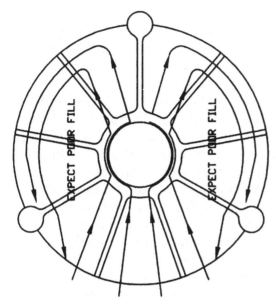

BACKFILL THEORY OF METAL FLOW

Figure 1

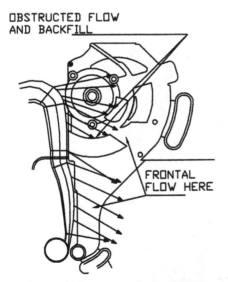

OBSTRUCTED FLOW
AND BACKFILL

FRONTAL
FLOW HERE

FULL FRONTAL FILL PATTERN

Figure 2

An example of typical behavior of the super heated liquid casting alloy during cavity fill confirms the theory. It is depicted in Fig. 3. A short shot, in which an insufficient volume of liquid metal is purposely injected, is shown to describe this phenomenon.

The difficulty in die casting a round shape is certainly evident here. This metal feed strategy concentrates the fill pattern on the center so that the metal streams will not circle the perimeter first and back fill the center where it is impossible to vent out entrapped air and gases. Notice how the center hub of this cast shape interrupts the metal stream to make it more difficult for the metal to reach the top edge of the cavity.

This casting was created by a short shot so that the cavity purposely did not fill completely. There can be no doubt as to the location of the gate inlet because of the almost acceptable condition of that zone of the cast shape. The last places to fill are clearly emphasized by the complete absence of metal in those zones. It is evident that there are too many overflows and they are misplaced. Obviously, the purpose of the short shot was to identify the root cause for poor fill as part of a Six Sigma exercise.

Cavity prefill is a strategy that North American die casters sometimes copy from a practice more common in Europe.

Figure 3

It is a technique that fills up to 35% of the cavity volume before the fast shot is initiated. This partial filling during the slow shot phase of the process helps to overcome defects caused by too much turbulence at details of the cavity that are close to the gate location and present severe obstructions to flow.

This logic is usually used after the metal feed system has been designed and tried out in an attempt to minimize casting defects.

Gate speed is another of the several conditions that determine what happens. This event is best presented by the analogy to the adjustable nozzle of a common garden hose as displayed in Fig. 4. Like the hose, the stream of liquid metal feeding into a die casting die can vary from coarse particles or droplets, through continuous jet flow, and ultimately to a finely atomized mist.

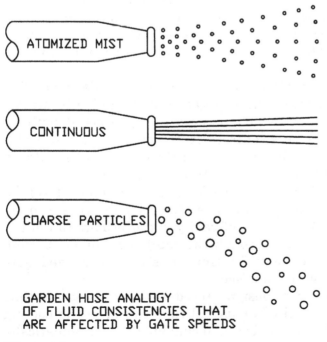

GARDEN HOSE ANALOGY
OF FLUID CONSISTENCIES THAT
ARE AFFECTED BY GATE SPEEDS

Figure 4

The atomized mist describes the gate speed range that produces the highest quality zinc die castings. The chemical and physical properties of the casting alloy that is to be injected into the die cavity determine which fluid consistency is best.

The recommended speed for aluminum alloys is in the range of 85–130 ft/sec (1020–1560 in./sec), and recommended gate thickness is approximately 50% of the wall thickness of the casting adjacent to the gate.

Magnesium alloys are produced with thinner gates and at speeds that vary from 150 to 200 ft/sec (1800 to 2400 in./sec).

Zinc alloys are produced with very thin gates between 0.012 and 0.020 in. (0.3–0.5 mm) and gate speeds from 170 to 200 ft/sec (2040 to 2400 in./sec) are suggested.

At first glance, it would appear that casting quality may be enhanced by an increase in the gate speed, but adverse consequences are encountered at excessive speeds. Aluminum, for example, is extremely abrasive (sand paper is made with aluminum oxide) and also has an affinity to absorb iron. Gate speeds in aluminum above 130 ft/sec (1560 in./sec) cause serious erosion of the die steel.

This is an insidious condition that has a delayed effect upon casting quality. As the gate orifice gradually erodes, the area increases, which reduces the gate speed with no change in the process control. Excessive gate speed can also cause too much turbulence within the cavity that results in vortices or swirls and voids in the casting surface.

Gate speeds can be calculated by the simple formula: $A_g = V/tV_g$, where A_g = gate area (length × thickness) in sq. in. or sq. mm; V = volume of cavity and overflows (all of the metal that must pass through the gate) in cu. in. or cu. mm; t = cavity and overflow fill time in sec; V_g = allowed gate velocity in in./sec or mm/sec.

While most gate design procedures used today are time based, since control of cavity fill time can keep the casting alloy liquid, the most effective technique is to mathematically define the casting. It is divided into zones of fill that either receive liquid metal directly from the gate, or are backfilled after the gated zone has been filled.

This logic examines the volume-to-surface-area ratio to establish the freezing rate of the casting alloy during cavity fill. The higher this ratio is, the quicker the freezing rate will be.

This method also requires the distance that the liquid metal must travel from where it enters the cavity at the gate to the farthest extremity of the casting from the gate, a number that also drastically affects the solidification schedule of the casting alloy. These data are easily determined by merely measuring the distance. Many experienced die casting engineers just use a piece of string for this purpose.

This strategy naturally leads to locating the last point in the casting to be filled with metal, which is critical to casting quality. It calls for identification of class "A" surfaces and then designing the cavity fill pattern so that the last place to fill is not within such a critical area. More will be offered on this important subject when vent designs are discussed in other chapters.

Computer aided engineering software is commercially available and in fairly common use in the generation of die casting firms that have survived the severe economic recession in manufacturing at the start of third millennium. Only one, however, has been developed solely for the high pressure die casting process. It is used to create runners in a three-dimensional wire frame that can be surfaced or converted to a solid model for subsequent CNC machining into the die steels. Mathematical cavity fill calculations are used to define the shape to be cast and the optimum process parameters to produce it. It requires considerable experience and knowledge of die casting for effective first shot success, but this author prefers it to others because it is simple and logical to use. The results are good in cases where minute (<0.02 inch diameter) porosity is acceptable.

The other software programs are based upon either finite element or finite difference analysis methods and are essentially three-dimensional simulators. The problem is that they are restricted by volume of fluid (VOF) technology that requires stationary grids. This limits their ability to deal with the highly dynamic turbulent flows experienced in high pressure die casting. Thus, they only provide some insight when an engineer is not very familiar with the process and are not substitutes for experience.

Unfortunately, appealing color and animation seem to be critical to marketing simulation software to the die casting industry, rather than strategic results.

Smooth particle hydrodynamics (SPH) appears to be the ultimate simulator to address porosity requirements that are expressed in parts per million opportunities (ppm) to produce castings. SPH has been in development for several years to streamline algorithms so that calculation time requirements can be reduced from weeks, to days, to hours, to minutes. The course of this work could take two different paths that depend upon the health of the die casting industry. It could be used as a research tool for solving flow problems for large captive die casting operations such as the car companies, or it could develop into a commercially available software.

7

Metal Feed System

Runners and gates are the usual name for the metal feed system in die casting nomenclature. This description, however, takes away from the real function, which is to convey liquid metal via the runner, through the gate, and into the cavity and overflow. An important event that is part of the feed system is the venting of air in the cavity in front of the metal stream that has to be exhausted out into the atmosphere for acceptable casting quality. Otherwise, the air will compress and form a back pressure that hinders full cavity fill.

View the metal feed system as a pump and conduit for the casting alloy between the source (holding furnace or crucible) and the die cavity. This may be over simplifying the die casting process, but it need not be complicated...it is merely a plumbing system. Super heated liquid casting alloy has a viscosity similar to water and therefore behaves as a hydraulic fluid. All of the scientific rules like "water does not flow uphill" apply.

The flow is turbulent in high pressure die casting to minimize cavity fill time. All of the benefits and drawbacks must be understood and accepted. The liquid metal stream can be expected to follow a straight line and will be broken up by any barrier it hits, which disrupts the pattern. Liquid metal streams do not change direction easily so sharp bends in runners need to be avoided. Remember, however, that the turbulent stream can be directed so it is logical to aim it at the most critical detail in the cavity if obstructions can be avoided.

The complication comes in with the peripheral equipment like pumps, electronics, computers, robots, safety guards, etc. The metal feed system is buried deep within the die casting cell and out of view during the casting cycle. Therefore, all strategies have to be exercised in the engineering stage.

Geography is a word sometimes used to describe the amount of space on the die layout for the metal feed system. It is important to avoid abrupt changes in direction, which many times requires space. As discussed in Chapter 6, too often the size of the die casting die is intentionally made small to fit into a certain size machine or to minimize die material costs. All of the wrong reasons limit opportunities for smooth unrestricted flow through the metal feed system. This is why the metal feed system should be designed before the die is laid out.

Gravity feed is to be avoided in the higher temperature alloys of aluminum and magnesium that are cast by the cold chamber process. Like water, these casting alloys will fall by gravity if allowed to be gated down into the cavity. This slow gravity feed, occurring before turbulent high speed and high pressure start to fill the cavity, introduces a separate and colder portion of casting alloy that will solidify earlier than the main stream. Cold shut and porosity defects can be expected.

In the cold chamber process, the constant velocity system explained in Chapter 3 is normally used in North America. This calls for the cold chamber, sometimes called the shot sleeve, be completely filled with a volume of liquid metal equal to the volume of the runner system, cavity and overflows, plus a biscuit. During the slow shot phase of the casting cycle, the air in the chamber is theoretically displaced with liquid casting alloy.

A biscuit thickness of at least 1 in. assures that the total system will receive metal in addition to providing enough metal supply to pack the cavity during the intensification phase. Without a respectable biscuit, the possibility of non-fill in the cavity due to a lack of metal injected into the feed system is frightening.

Minimize the incidence for porosity caused by entrapped air by recognizing the critical aspect of the fill level of liquid metal in the cold chamber after pouring and prior to initiating the slow shot phase of the cycle. The shot end of the casting machine is designed with a minimum plunger displacement to accommodate the stationary platen thickness and also provide for a reasonable cover die thickness and die shut height.

Therefore, the empty volume of the cold chamber is the product of the area of the cold chamber times the displacement. Only very massive castings such as automotive transmission cases require enough metal to fill the cold chamber. Most die castings require only enough metal to fill 25–40%, which leaves the balance of the air in the shot chamber to be exhausted through the metal feed system. This condition must be dealt with carefully and will be expanded upon later.

One method of eliminating this situation that keeps coming up is the use of a vacuum suction of the liquid metal from the holding bath to the cold chamber without a pour hole. Another is the vertical die casting machine.

The hot chamber process used for zinc and some magnesium operations does not experience the air entrapment problem, but proper design of the sprue post (spreader pin) and bushing must be considered.

Two types of sprues are used, the constant area sprue system and the runner sprue, which are illustrated in Figs. 1 and 2.

The constant area sprue theoretically provides a constant area for liquid metal that exceeds the runner area by reducing the thickness of the space between the male post and the female bushing as the diameter increases. Caution should be exercised to assure that the constant area continues at the base of the post and the intersection with the runner. A major problem with the standard style of sprue post shown in Fig. 1 is the opening around the circumference of the post

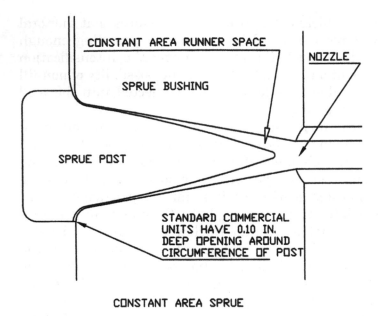

CONSTANT AREA SPRUE

Figure 1

(sometimes called the spreader pin) that is necessary to provide an adequate conduit to the runner. It is only coincidental that the area of this element has the same cross-sectional area of the rest of the metal feed system and the nozzle.

The runner sprue, on the other hand, provides no theoretical space between post and bushing, but a constantly decreasing runner area is machined into the post between the nozzle and the runner in the die parting plane. This configuration is depicted by Fig. 2.

It should be obvious that the cross-sectional areas of the runners can be better controlled with the runner sprue design.

Of course, the runner from the sprue to the cavity is just an extension of the sprue runner, but it is important that the runner area constantly decreases in area until it reaches the gate (Fig. 2). Again, the hydraulic formula of $Q = AV$ is utilized.

As a matter of fact, the best way to design the runner is to start at the gate and work in the reverse direction to the flow of the casting alloy back to the metal source. The runner area should be increased by a factor of 5% at each bend or

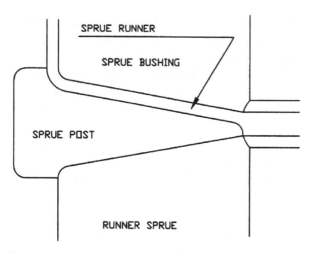

Figure 2

split, and another 5% at the intersection of the runner to the sprue.

This strategy is also used in the design of runners for the cold chamber process as shown in Fig. 3.

It is also referred to as a sprue runner and is intended to minimize the effect of the phenomenon called vena contracta

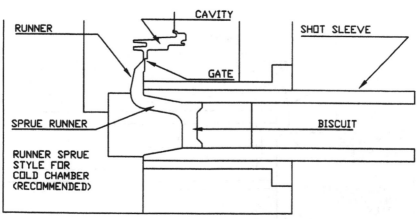

SECTION THROUGH COLD CHAMBER FEED SYSTEM

Figure 3

to reduce air entrapment as the liquid metal makes a 90°
change in direction from horizontal to vertical. Remember,
at super heated temperatures, the casting alloy behaves like
a hydraulic fluid and wants to follow the path of least resistance. Thus, every effort to streamline the metal feed path
will be rewarded with an improvement in internal integrity
of the product.

The usual style where the directional change is made
abruptly at the intersection of the biscuit and runner is
described in, Fig 4, and the possibility for air entrapment in
this option is obvious.

The shot sleeve is a constant problem to most die casters
since it wears away near the pouring hole and requires
continuous lubrication. The shot sleeve is a steel tube that
functions in an open air environment. It is usually water
cooled at the tip end. Super heated casting alloy is poured
into the other end and runs along the length, transferring
heat until it reaches the front end.

The liquid metal is then allowed to stabilize before the
plunger tip starts to move. During this time, a skin of solidified metal forms against the inside wall, especially near the
water cooled end.

As the metal fills the sleeve and starts to move into the
runner system, the skin is peeled off the wall by the tip. These

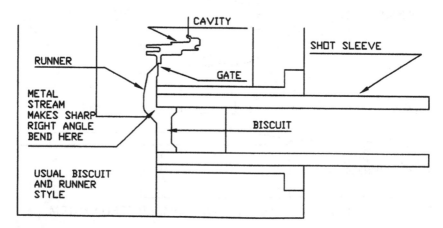

Figure 4

flakes are then entrained into the liquid metal stream and travel to the gate—the first restriction where they get trapped and cause a partial blockage. This restriction is signaled by streak marks in the casting that emanate from the gate because the flow though the gate is a variable. Random defects such as cold shut in areas remote from the gate.

Normally, this condition can be eliminated by increasing the metal temperature or changing the slow shot plunger velocity.

Increasing metal temperature reduces the thickness and strength of the flakes so that they may remelt before they reach the gate restriction.

Increasing slow shot plunger velocity allows less time for the skin to form, and the flakes that do are thinner and shorter in length.

Reducing the stabilizing time between pouring and the start of the first shot stage also produces thinner flakes.

The key to controlling this problem is in maintaining a reasonably constant shot sleeve temperature. Further research is needed to determine the ideal shot sleeve temperature, but early trials suggest a range between 400°F and 480°F.

The runner is immediately down stream from the source of the casting alloy and serves as a conduit between the metal supply at the biscuit for the cold chamber process or the nozzle for the hot chamber process. Since the liquid metal follows the path of least resistance, abrupt changes in direction should be avoided or provided for in the design of the metal feed system. Separations of the main runner into separate branches are minimized because splashing and air entrapment occur at each junction.

The runner design has to be as streamlined as possible to literally provided a path of least resistance. Another important consideration is the speed of the liquid metal as it travels through it. This is done by constantly reducing the cross-sectional area as each change of direction or impediment to flow is encountered. It is recommended that the area is reduced 5% at each directional change or split. The runner area has to be at least as large as the cross-sectional area of the gate so that the flow rate does not decrease as the metal passes through the gate. It is more preferable, however, to design the runner element that directly feeds the gate with 10–20% larger area. When tapered tangential runners are used, this also is a major factor in defining the flow angle.

The perfect cross-section for a runner is the circle since it offers the most thermal efficiency. It is important to hold the heat loss as low as possible. However, since the runner is formed by the two die halves, it is more economical to machine it into one half and then the other half merely forms a flat side when the dies close. Draft must be included at each side, so that most common cross-section is the trapezoid. The trapezoidal shape should be designed as close to a square as possible to contain the heat in the metal.

Most runners, however, are designed wide and shallow for quicker cooling after the cavity has been filled and the shot is in the dwell stage during solidification. Experience has proved, however, that if the depth of the runner does not exceed 1 in. except in extremely large shots, adequate cooling can be provided. Figure 5 illustrates both runner cross-sections discussed here.

The ideal runner system is balanced for multiple cavity dies in which the liquid metal reaches the gates into each die cavity at the same time. Then, each cavity will fill *in* the same time and *at* the same time. The importance of this basic strategy is that uniform quality will be produced. If this policy is followed, there is no logical reason to sort cavities to

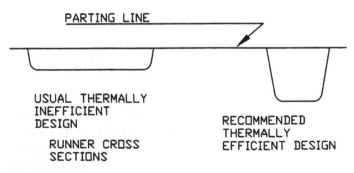

PARTING LINE

USUAL THERMALLY
INEFFICIENT
DESIGN

RECOMMENDED
THERMALLY
EFFICIENT DESIGN

RUNNER CROSS
SECTIONS

Figure 5

separate one quality level from another. If one part is bad, they all should be, and conversely, if one is good, they all will be.

Seek a more thermally efficient section so that the more of the pouring heat is retained by the casting alloy on its path from the shot tip to the gate. Yes, a square section with draft on each side is more effective, and an easy way to develop this runner shape is to calculate the square root of the area to determine the depth, but remember to limit it to 1 in. except for very large shots.

Why then do the poorly designed runners produce acceptable castings? Again, it must be understood that the die casting process is extremely forgiving. The trouble comes in when too many rules are broken and complex casting problems result.

In conclusion, design runners should constantly decrease in cross-sectional area from the metal source to the gate so that the velocity of the liquid metal constantly increases as it travels through this conduit. This is critical because any deviation in this velocity pattern will generate turbulence and trap air. This air is encapsulated in an envelope of liquid metal which, of course, cannot be vented out of the cavity without spewing metal into the ambient environment. The result is gas porosity in the casting that is a major reason for rejection. The metal must exit the runner system in this manner through a gate to enter the cavity.

Gate design which includes the location, style, direction, and cross-sectional area is critical to effective cavity fill conditions that have a profound effect upon casting quality.

While there are many styles for gate design, the main three will be discussed. A poor design of the popular *fan gate* is described here. Remember, the principle of constant reduction in area still must be considered for a proper fan gate design.

In Fig. 6 the mid area increases the area just before the metal reaches the gate, which slows down the flow velocity and then speeds it up, causing turbulence that entraps air.

The straight sides are easy to machine into the die steels and the tooling cost is low, but the negative effect on casting

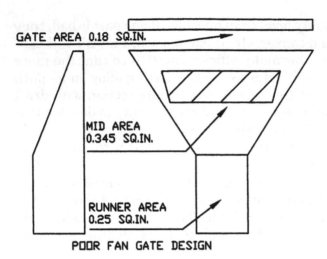

GATE AREA 0.18 SQ.IN.

MID AREA
0.345 SQ.IN.

RUNNER AREA
0.25 SQ.IN.

POOR FAN GATE DESIGN

Figure 6

quality will increase production costs of the casting that far exceed the lower tool cost.

The graphic in Fig. 7 describes an improvement in the fan gate design since the mid area falls within the prescribed limits. The concave side walls reduce the horizontal dimension so the mid area is reduced.

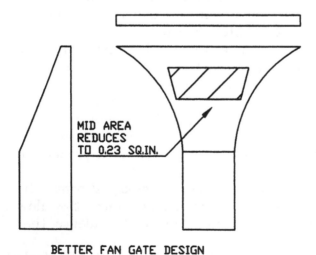

MID AREA
REDUCES
TO 0.23 SQ.IN.

BETTER FAN GATE DESIGN

Figure 7

Figure 8 illustrates the best style for a fan gate where the mid area is an average of the runner area and the gate area. Of course, this further increases the machining cost, so most die casters settle for the straight sided design. However, since the advent of CNC tool path programming, the cost factor is rapidly disappearing.

In considering the fan gate style, it must be understood that the gate speed varies from very fast in the center to almost zero at the ends. The speed variance is difficult to calculate, so most die casters merely use the hydraulic formula of $Q = AV$ to calculate the average gate speed.

This procedure is dangerous because the speed at the center of the fan can cause early erosion of the die steel if it exceeds 150 ft/sec. Typically, this gate speed variance generates vortices at either side of the center and swirls are formed that result in poor surface finish and can be a site for gas porosity. This scenario is illustrated in Fig. 9 with gate vectors, the length of which describes gate speed.

The distance between the gate exit from the runner and the leading edge of the cavity is called the gate land which is a function of the gate thickness, but usually is limited to a

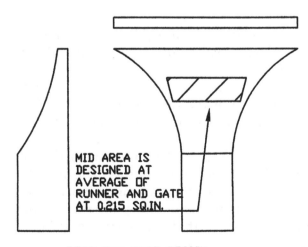

MID AREA IS DESIGNED AT AVERAGE OF RUNNER AND GATE AT 0.215 SQ.IN.

BEST FAN GATE DESIGN

Figure 8

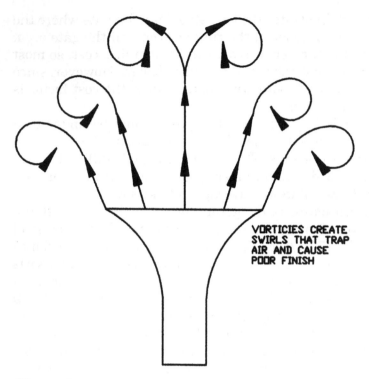

VORTICIES CREATE
SWIRLS THAT TRAP
AIR AND CAUSE
POOR FINISH

Figure 9

maximum of 0.10 in. If the land is greater, there is a danger of
the gate freezing enough to effect the movement of metal during the intensification shot phase.

Again, the reason that the fan gate works in so many
cases is the forgiving nature of the die casting process. The
gate speed is not constant and it is difficult to control the fill
pattern.

The *chisel gate* is used mostly to feed remote portions of
the cavity where help is needed to strengthen the feed for
surface and internal integrity of the casting. Like the fan
gate, the speed varies and is faster at the center. However,
the width is usually so narrow that the swirling effect is
greatly diminished.

Such a helper runner and gate is a useful adjunct to the
main metal feed system. Where the main gates are chisel

GATE AREA

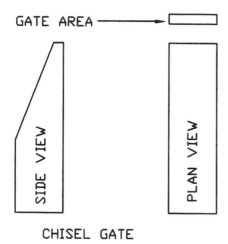

CHISEL GATE

Figure 10

gates, however, the fill pattern is very limited and the casting quality suffers. Figure 10 describes a typical chisel type gate.

The *tapered tangential runner*, sometimes called the Australian gating system, deals with the undesirable condition in the other gate styles. A typical gate configuration of this style is depicted in Fig. 11.

The vectors represent the direction of flow and the length of each vector describes the relative gate speed. Since these vectors are approximately the same length, the gate speeds are considered to be constant.

The cross-sectional areas are computer designed to yield constant gate speeds. The flow angle of the metal that exits each runner element is a function of the inlet area to the gate area. This relationship is not linear so several cross-sectional areas must be calculated for each runner element in a manner that will achieve a constant flow angle and a constant gate speed.

This runner style may be designed manually or with a conventional CAD program, but the most accurate method is with the software known as Castflow®.

The gate is the final restriction upon the metal feed system and it has a profound effect on cavity fill time, gate speed, the shot end settings of the die casting machine, and, of course, the quality of the castings produced.

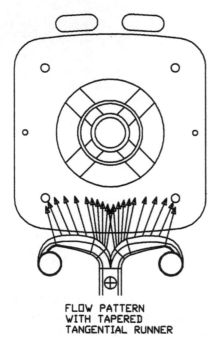

FLOW PATTERN
WITH TAPERED
TANGENTIAL RUNNER

Figure 11

Keep in mind that the casting alloy loses heat all through the cavity filling process, so it is minimized by increasing the gate area and maintaining the same gate speed. This is usually accomplished by increasing the thickness.

Caution must be taken, however, since enlarging the gate area also slows down the gate speed. This can be overcome by increasing the fast shot plunger velocity.

The runner area must also be large enough to support a larger gate area. Thus, these gate options should be considered prior to machining the runner into the die steels. Too many times, gates are revised to deal with quality issues while the die is in the casting machine. Often, the process should be examined rather than the die itself.

If you think about it, logic would suggest that even a scrap figure as high as 50% means that half of the castings produced meet the quality requirements. The die steels, including the gates, do not change between cycles! The process variables can and do experience deviations, however.

Another thing that must be addressed when discussing the metal feed system is that abrupt directional changes are not kind to the die casting process. The runner system is indeed a plumbing system in that it acts as a conduit for the liquid metal.

Yes, the metal behaves like a hydraulic fluid at temperatures above the liquidus since it has a viscosity about like water. Also, in die casting, we are dealing with extremely high pressures and turbulent velocities. At these high pressures and speeds, liquid metal resists any change in direction; kinetic energy is released and causes severe damage to the expensive die.

Gate area has profound effect upon casting defects and quality. The chart in Table 1 describes conditions in a qualitative manner that suggests what any gate area should be to accomplish a particular purpose (Von Tachach).

Gate area has no effect upon shrinkage porosity that is a function of volumetric shrinkage during solidification.

Generally, the higher the specific gravity of the casting alloy is, the more energy it releases at sharp directional changes. Thus, lighter alloys such as magnesium and aluminum can handle directional changes easier that heavier alloys like zinc, brass, and lead.

This also leads to cavitation which can badly damage a die surface in just a few shots.

Table 1

EFFECT OF GATE AREA ON QUALITY CHARACTERISTICS

SMALL GATE AREA		LARGE GATE AREA
HIGH GATE VELOCITY		LOW GATE VELOCITY
LESS ◄——————— GAS POROSITY ———————►		MORE [LARGER PORES]
BETTER ◄——————— SURFACE FINISH ———————►		WORSE
LIKELY ◄——————— SOLDER ———————►		LESS
WORSE ◄——————— EROSION ———————►		LESS
BETTER ◄——————— MECHANICAL PROPERTIES ———————►		WORSE
MORE CRITICAL ◄——————— SHOT END CONTROL ———————►		LESS CRITICAL
HIGHER ◄——————— INJECTION PRESSURE ———————►		LOWER
NOT APPLICABLE ◄——————— DIRECTIONAL SOLIDIFICATION ———————►		EVENTUALLY POSSIBLE
ESSENTIAL FOR THIN WALL CASTINGS ◄——————— WALL THICKNESS ———————►		FOR SIMPLE OR THICK CASTINGS

Cavitation is a phenomenon that escalates with liquid density of the casting alloy. An analogy to a large semi-truck making a turn at high speed to a small automobile in the same posture may be helpful to understand cavitation. Of course, it is more difficult to turn the truck since it is more massive and wants to continue in a straight line more than the car.

Cavitation is the generation of cavities in a fluid that occurs when local pressure falls below the vapor pressure of the fluid whenever bubble nuclei are present. A bubble carried along in a stream of liquid metal is not stable since local velocity and pressure are continually changing (NADCA, 1991). Bubbles normally collapse after a short lifetime. Often they collapse near the die surface. This is called an implosion and frequent repetition at the same spot can cause serious die pitting that usually occurs down stream (in the cavity) from the source of the bubble like a sharp bend (in the runner). Figure 12 illustrates this condition.

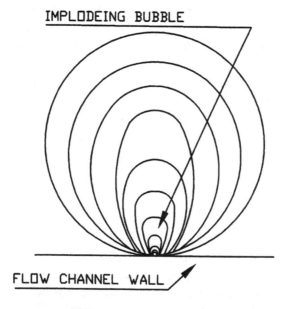

IMPLODEING BUBBLE

FLOW CHANNEL WALL

THE COLLAPSE OF A CAVITATION BUBBLE TOUCHING THE WALL OF A FLOW CHANNEL

Figure 12

Many times die casters are surprised by this die pitting when it occurs in zinc die castings because this material is considered more gentle to die steel surface. The explanation is that zinc is heavier and therefore resists any change in direction more than aluminum, which is much harder on the die steel.

Overflows are part of the metal feed system and serve several purposes. The primary reason for them is that they act as heat sinks and are normally located adjacent to the last location in the cavity to receive metal, which is the coldest in the system and where the incidence for a cold shut defect is strong. In this case, the overflow is designed with as much volume and as little surface area as possible.

Sometimes ejector pin marks are undesirable on a particular surface of the shape to be cast. This can be overcome by locating the ejector pin outside of the cavity on an overflow. Such an overflow is called a false ejector.

Due to the tendency for super heated liquid alloy to back fill, an overflow can be placed off of the last region to fill where partially solidified metal, excess air, or gas can be drawn away from the casting. The science behind this theory is debatable, however.

A strategically located overflow can be a good place for initiating an air vent.

Venting of the die cavity is almost as critical as proper gate design since entrapped air is a major cause of porosity. Air that may be become entrapped in a die casting comes from many different sources. Air occupying the shot sleeve and cavity prior to the injection of metal is the main source. Lubricant decomposition as a result of contact with hot metal also creates gas that must be vented out from the cavity prior to cavity fill to minimize porosity.

Vent area is defined as the smallest area of the vent as it reaches the outside edge of the die retainer where the air escapes into the atmosphere. It really is a function of the volume of the cavity, but since the gate area is also related to cavity volume, vents are normally sized by relating them to the gate area. Roughly, the vent should be 10–20% of the gate area, but most die parting planes are not large enough to accommodate more than 10% so this becomes the maximum that is mechanically possible.

The vent path is designed to exit the cavity or adjacent overflow at a thickness of 1/64 to 1/32 in., usually in the ejector die and then to step down to 0.007 (aluminum) to 0.004 in. thickness (zinc) at the boundary between the cavity insert and the die retainer. The thicker portion will fill with metal, which will freeze as it enters the thin part so that only air is exhausted into the room. The thick part is in the ejector die so that it can be ejected with the shot.

Strategy for the shape of the vent path is the reverse of that used for the runner: Sharp abrupt changes of direction are used to encourage the rapidly solidifying metal to freeze and to eliminate the possibility of squirting hot liquid metal out into the environment of the casting cell.

Gas and air movement is initiated by pressure build up in the die cavity. The air or gas velocity is measured by mach number or the velocity relative to the speed of sound. It is usually considered that the ideal speed is just below or at mach 1. The speed of sound in air can be calculated approximately by the formula

$$V_s = 331.4 + 0.6\,T_c\,\text{m/sec}$$

where T_c is temperature in Celsius, so that at an air temperature of 260°C, the speed of sound is

$$V_s = 331.4 + 0.6[260] = 487.4\,\text{m/sec}$$

In English units, at an air temperature of 500°F, the speed of sound is 19,284 in./sec.

Some research has determined that most of the air in the feed system is exhausted during the slow shot phase and very little escapes after the plunger velocity reaches 50 in./sec (Mangalick, 1976). Fill time can be related to the vent capacity of the die so that by increasing vent area, the cavity fill time decreases because back pressure is diminished.

Cavity fill time is integral to logical vent design because it is a function of gate area and gate speed. In other words, if gate area is increased to reduced cavity fill time without exceeding guidelines for gate speed, vent area also must be increased proportionally.

In addition, air in the cavity escapes through core slides and ejectors pins. Die blow also contributes to air loss during cavity fill. If it is assumed that the ideal speed at which air should be exhausted in front of the metal stream is 0.8 times the speed of sound, or 15,427.2 in./sec, the optimum vent area can be calculated. Other reasonable assumptions that can be made are that the air is lost from the runner during slow shot, and that 60% of the air in the cavity is lost through leakage and die blow.

A formula to calculate optimum vent area is

$$V_a = 0.5\,V_c/C_{ft}[V_v]$$

where

V_a is vent area;
V_c is cavity volume;
C_{ft} is cavity fill time;
V_v is vent velocity.

An example is offered here. Where cavity volume is $10\,\text{in.}^3$ and zinc is the casting alloy, cavity fill time is $0.02\,\text{sec}$, then

$$V_a = 0.4\,(10)\,\text{in.}^3/0.02\,\text{sec}\,(15,427\,\text{in./sec}) = 0.013\,\text{in.}^2$$

Since a vent thickness of 0.004 in. is recommended for zinc, the vent area of $0.016\,\text{in.}^2$ can be designed as 0.004×3.25 in.

A device called a massive chill plug is sometimes used to rapidly freeze the vent off with water cooling since the only purpose of the vent is to remove air, not to transport metal.

So far, a natural venting system has been described here, but there are vacuum systems that partially evacuate the metal feed system and cavity that are sometimes more effective than the natural vents. These are referred to as power vents and care must be taken to get tight fits in the die in locations like the parting planes, ejector pins, and core slides so that the vacuum cannot bring in air from the atmosphere.

With either type, however, the strategic location of the vents at the last place in the cavity that is expected to receive metal is more critical than the sizing. If this decision is not made correctly, and liquid metal enters the vent before the cavity is filled, it will block off the vent, which will stop any further venting.

DESIGNING THE FILL PATTERN

It is possible to roughly design the fill pattern when conventional runners and gates are used by directing or aiming the turbulent metal stream at the critical region of the shape to be cast. After this rough strategy has been established, there are techniques available to calculate the fill pattern more accurately.

The layflat technique describes a manual method to define the fill pattern. The shape to be cast is laid out flat as if it were to be made from sheet metal. After the gate location has been defined and the flow angle determined, mathematical calculations are made to define volume, surface area, and distance that the liquid metal must flow through the gate to fill the cavity.

Flow simulation software, into which a three-dimensional model is imported, the gate and runner attached, and the casting process defined modeled that can graphically simulate the flow pattern more accurately. It is wise to analyze the flow pattern before accepting a final design of the metal system for all shapes with any degree of complication. This needs to be done before the feed system is cut into the die steels.

Tapered tangential runners, when properly designed, direct the flow of metal and control its speed as it exits the runner through the gate. Calculations are accurate and the results are predictable. It should be noted, however, that this author sees too many poorly designed tangential runners at too many die casting plants. In view of its potential for this purpose, we will focus upon this type of metal feed system.

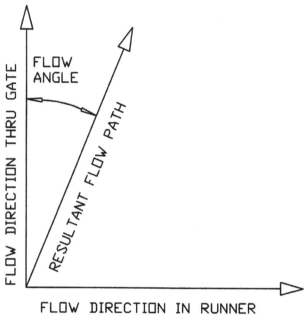

Figure 13

We will start with the flow of liquid metal in the runner and its predictability that it will follow the path of least resistance in the direction of the runner. This is represented by the horizontal vector in Fig. 13.

When the runner is located tangential to the edge of the cavity and a gate orifice is cut between the runner and the cavity, the flow path also wants to exit the runner at a right angle, or in the vertical direction. However, it has an equal interest in continuing the horizontal flow. The vectors shown, by their relative length, quantify the gate speeds. Since the vertical vector, normal to the horizontal line, is shorter, the normal gate speed is slower than the resultant or true gate speed.

What actually happens is that the metal stream will follow a path in the direction of the resultant of the vertical and horizontal vectors. This is called the flow angle. It is the basis for accurately designing the fill pattern that is bounded by the edges of the metal stream that follows the resultant vector

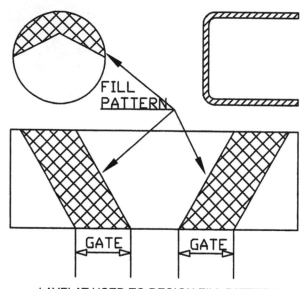

LAYFLAT USED TO DESIGN FILL PATTERN

Figure 14

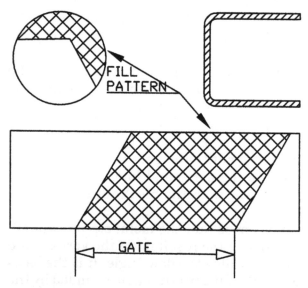

Figure 15

until it meets an obstruction, such as a core or interrupting wall within the cavity.

These edges are three dimensional and can be traced upon a prototype model or CAD description of the shape to be die cast. The simple, but difficult to fill, hat shape with the flow pattern traced onto it is shown in Figs. 14 and 15 in lay flat form. This old but effective method is used to visualize the hat shape fill pattern. This pattern would be accomplished with a double tapered tangential runner system and is not necessarily the optimum fill strategy. It is used only to explain this design technique.

The hat shape was discussed earlier because it is so difficult to fill without defects. The metal streams run around the skirt with a direct frontal flow, but only back fill the top with colder metal. Cold shut or lamination defects usually occur because the metal streams flow around the outside and not over the top. In the case shown in Fig. 14, a small portion of the top is filled, however.

It is always desirable, where mechanically possible, to explore other gate strategies.

Therefore, a plan to swirl the liquid metal into the skirt sometimes is more effective for the hat shape. The idea here is to continue the frontal fill pattern as long as possible. Figure 15 shows that some of the metal stream also fills a portion of the top. This is a more desirable option because the fill pattern covers a larger area, and venting of the top is possible.

8

Process Control

Control of the process begins with the die casting machine, especially the vicissitudes of the shot end. "Vicissitudes" is the word used because even relatively new machines stray from the original base line design data in proportion to the amount of abuse they receive in production. The key fixed elements of the die casting machine are: the diameter of the shot cylinder, rod diameter, plunger displacement range, maximum fast shot plunger velocity, locking force intensification ratio, and rise time to intensification.

Variables to be addressed in the cold chamber process are the plunger tip diameter, inside diameter of the cold chamber, accumulator pressure, operating metal pressure, biscuit thickness, gate speeds, cavity fill time, and metal temperature at the end of cavity fill. In the hot chamber process, the goose neck replaces the cold chamber, and the nozzle length and outlet bore are added.

Locking force is rarely a limiting factor because platen sizes are small compared to the clamping tonnage, and size not force usually determines which machine to run a particular die.

Thus, the die casting machine provides the pump at the shot end to supply the super heated casting alloy to the die, and the clamp to hold the die halves shut. Compatibility in all areas between the machine and the die is important to efficient productivity and quality.

The details of the metal feed system discussed in the prior chapter define how the process variables must be controlled for statistical results, designed to eliminate sorting good castings from bad ones. Predictable results of salable product can be expected in at least the 98 percentile for aluminum and magnesium die castings, and at least the 95 percentile for the most challenging product, cosmetic zinc castings.

The role of intensification is covered here since it is universally used in cold chamber operations to minimize gas porosity. It is defined as the controlled increase of pressure on the casting alloy at the end of cavity fill, immediately following impact (McClintic, 1995). It is accomplished by increasing the hydraulic pressure above nominal by shifting to alternate relief valves, opening high pressure accumulators, or operating multipliers called cylinder intensifiers. The usual ratio is 3:1 compared to the pressure used for filling the cavity.

Intensification is initiated by a position-based signal during the deceleration near impact. Biscuit thickness consistency becomes critical to the timing of this final squeeze. Since the objective is to compress gas porosity voids that have occurred during cavity fill, the strategy is to continue to squeeze metal through the gate orifice before it solidifies to form a denser cast shape.

An impact pressure surge or spike takes place briefly (25/100,000 of a second) at the end of cavity fill and the plunger abruptly decelerates. This is due to the inertia of the hydraulic fluid and the mechanical components of the shot system. Excessive flash is the result of too great a peak in impact pressure. Modern die casting machines are equipped with a deceleration feature on the shot end, triggered either by a limit switch or encoder signal, at the proper shot cylinder position. Managing deceleration impact and intensification pressures properly plays a big role in internal integrity and casting quality.

The shot system of the casting machine operates with a combination of hydraulic principles in addition to the pressure applied upon the casting alloy. High pressure is used to pack the liquid metal into the cavity, which occurs both during and after cavity fill. Velocity is the other critical principle that has a profound effect on the quality of castings produced. Effective management of both is what die casting process control is all about. Critical parameters need to be identified and their effect upon casting quality must be quantified before there is any real process control.

A typical shot trace that displays both pressure and velocity is described here. All process monitoring systems produce some form of this simple chart to illustrate this important information graphically. Such a chart quantifies current conditions that may then be compared to the actual quality being produced. A master trace may also be generated to graph acceptable limits of a designed process so that all casting cycles can be compared to this optimum.

This does not solve the problem however; it merely identifies it. From this point, knowledge of the die process and experience control or improve upon the existing casting quality.

The different stages are clearly illustrated in the graph shown in Fig. 1 that describes the performance of a typical

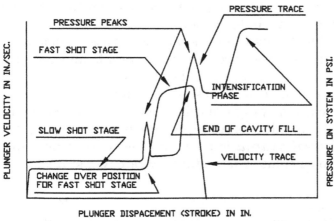

PLUNGER DISPACEMENT (STROKE) IN IN.
TYPICAL SHOT TRACE

Figure 1

cold chamber shot cycle. It is made in two stages of velocity where the shot sleeve and runner system is filled with a minimal amount of turbulence during the slow shot stage, which is usually in the 5–30 in./sec range. The cavity is filled during the fast shot stage, which normally is in a range of 80–150 in./sec. sec. Sometimes, the shot valves of cold chamber machines are configured to use a three-stage plan, where the first stage is very slow. The purpose is only to move the plunger tip forward enough to close the pour hole off without splashing metal out. In this case, the second stage performs as the slow shot described above.

To determine the proper changeover position, the volume (quantity) of metal required to fill the total shot, including casting, runner, biscuit, and overflows, must be calculated. Then, it is important to also determine the volume of all but the casting and overflows so that the distance traveled to bring the metal within an inch or two of the gate orifice can be calculated. This distance is the changeover position. (A helpful formula follows shortly in this discourse.) It can readily be observed that the plunger velocities are constant and acceleration occurs only after the changeover position is reached. This condition is called a constant velocity mode, which is practiced by most North American die casters.

In other geographical regions, especially Europe, a constant acceleration mode is practiced. There is some evidence that there is less incidence of gas porosity when this method is followed because turbulence is reduced.

A graph illustrates the constant acceleration shot mode in Fig. 2. Pressure peaks are lowered in profile and the gradual build up of velocity can also be observed.

Since die casting can be described as a turbulent process because of the extremely high velocities and time in milliseconds, the advantage of a reduction in turbulence becomes apparent.

The universal hydraulic formula, a very basic principle that is used to calculate the metal flow rate (Q), states:

$$
\begin{array}{ccc}
Q & A & V \\
\text{(quantity or metal} & = \quad \text{(area of} \quad \times & \text{(velocity of} \\
\text{flow rate)} & \text{plunger tip)} & \text{plunger)}
\end{array}
$$

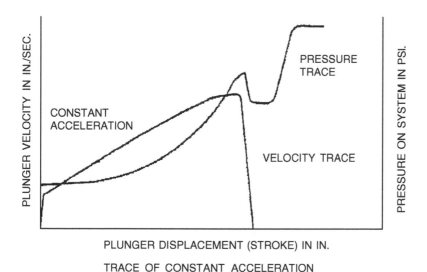

PLUNGER DISPLACEMENT (STROKE) IN IN.

TRACE OF CONSTANT ACCELERATION

Figure 2

For example, the area of a 3 in. diameter plunger tip is $\pi \times \text{radius}^2$, or $\pi \times 1.50^2$, or 7.065 sq. in. Then, if the fast shot velocity is, say 90 in. per sec,

$$Q = \pi \times 1.50 \, \text{in.}^2 \times 90 = 635.85 \, \text{in.}^3 \, \text{per sec}$$

It should be noted that the quantity may be increased or decreased by changing either the diameter of the tip or the velocity.

The velocity trace shown here describes acceleration and deceleration that relate to changes in resistance to flow as the die cavity is being filled.

Pressure on the shot system affects plunger velocity so it is also important to understand the equations that are used to calculate pressures, and especially the pressure that is applied to the casting alloy during cavity fill. Significant pressure points during the shot cycle are exagerated in Fig. 3. This all establishes the capability of the casting machine of choice to supply liquid metal to the die, through the runners and gates.

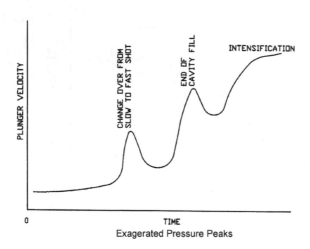

Figure 3

The operating pressure and flow rate must be calculated from measurements of the plunger velocity and pressure, which are usually obtained by instrumentation that quantifies the machine shot system. Figure 4 illustrates the method used to instrument a cold chamber machine to derive the

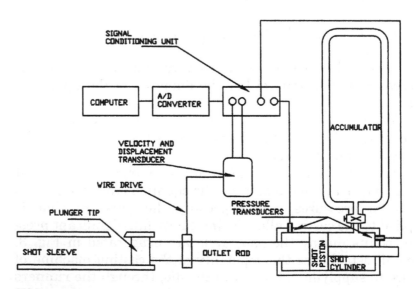

Figure 4

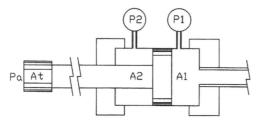

PRESSURE SCHEMATIC

Figure 5

metal pressure and flow rate. The hot chamber configuration may be instrumented in a similar manner.

The formula used to calculate the operating pressure applied to the casting alloy is

$$P_a = (P_1 A_1) - (P_2 A_2)/A_t$$

where:

P_a = pressure on casting alloy
P_1 = inlet or accumulator pressure
P_2 = exhaust pressure
A_1 = shot piston area at inlet
A_2 = rod diameter at outlet (usually piston area minus outlet rod area)
A_t = area of plunger tip.

Figure 5 illustrates a schematic description of the hydraulic shot cylinder and serves as a reference for the above formula.

Critical slow shot velocity (Vss) is the proper slow shot. Plunger velocity cannot be picked at random because of the phenomenon illustrated in Fig. 6. If it is too slow, a pocket

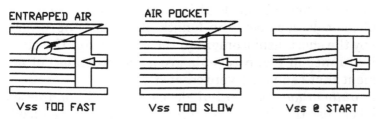

ENTRAPPED AIR AIR POCKET

Vss TOO FAST   Vss TOO SLOW   Vss @ START

Figure 6

Table 1 Critical Slow Shot Velocity

Cold chamber diameter (in.)	Vss (in./sec.)
1.50	16.9
2.00	19.5
2.50	21.8
2.75	22.9
3.00	23.9
3.50	25.8
4.00	27.6
4.50	29.3
5.00	30.9

of air will be entrapped just in front of the plunger tip. Then, as the metal stream changes from the horizontal to the vertical direction, the turbulence at this point encapsulates the air with an envelope of liquid metal, which can only end up in the casting as porosity. If too fast, a wave will break, which also entraps an air pocket with the same result.

The slow shot velocity for any die requires calculation in the computer aided engineering programs, or it may be chosen from Table 1, which is abridged here to show the most popular cold chamber sizes for fill levels of 30–40%. This table was developed by Dr. Lester Garber during wave celerity research work conducted at the University of Rhode Island and Prince Corporation on behalf of the Die Casting Research Foundation.

Air entrapment in the cold chamber is a major source of porosity in the casting by virtue of a low fill level of liquid metal just before the plunger tip starts to move forward and the slow shot velocity. To illustrate this phenomenon in real time, a short shot has been made for the photograph in Fig. 7. Note the long biscuit that is formed by the inside diameter of the cold chamber, the length of which is purposely created by the short displacement of the plunger tip. The close up of the runner in Fig. 8 reveals great voids because of air entrapment that cannot be vented away.

The automation vs. manual decision usually addresses direct labor cost that can be eliminated or at least reduced. This is true, of course, since the die casting operator can be

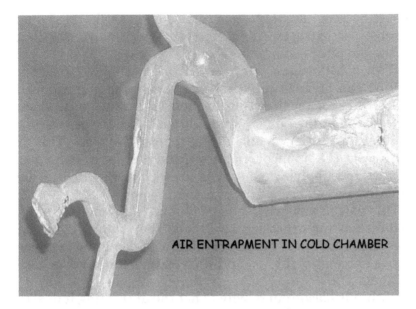

Figure 7

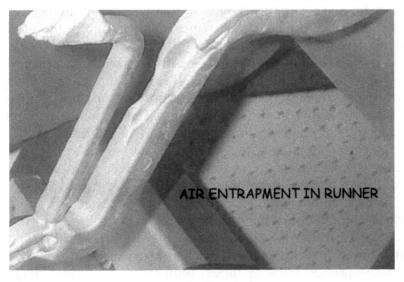

Figure 8

replaced by programmable controls to operate the machine; and robots or extractors and conveyors to remove the shot from the die after each cycle (Fig. 6).

With a little more commitment to excellence, the trim operation can be automated when the robot is programmed to place the shot on the trim die and then remove it. From here, it does not take much imagination to picture secondary machining operations being brought into the loop. This is called a work cell, in modern jargon, and there are companies dedicated to supplying this type of technology to the die casting industry.

This writer was privileged to observe a completely auto-mated die casting operation in Japan! They call it lights out die casting because the lights are turned off when the last per-son goes home at night ... and die castings are produced all night long.

With this background, why is not the die casting industry operating automatically with only a few managers and super-visors? Well, even though complete automation is possible, it requires near perfect conditions from fit tolerances to almost total repeatability. Thus far, this perfection has been lacking.

Computerized process controls have, however, taken a hold in die casting applications and there is a great deal of dif-ference between the consistency possible through automated controls and those left to humans.

The variables of metal temperature to die temperature to gate speed to cavity fill time to dwell time to duration of blow off and spray time profoundly affect product quality. There-fore, the substance of the following section will address this technology, for it is truly directed to quality assurance.

It is important that any automatic process control be pro-grammable, repeatable, and responsive (valves must activate in less than 3 msec) from shot to shot and set up to set up. With conventional manual control, there is too much guess work at the shot system. Acceleration and deceleration con-trol must be coupled with the more basic fundamentals. The control must react to and correct for external variations that affect the process such as metal temperature, plunger drag, load pressure, and the viscosity of the hydraulic fluid.

Process monitoring that was illustrated earlier describes a typical shot trace that is the graphical form that process monitoring takes. In addition, it is important that a proper monitor is capable of mathematical calculations to quantify the velocity, time, and pressure variables encountered.

There are many such systems available to monitor just one die casting machine or many machines. As a matter of fact, a central control room is usually set up in or near the die casting department. The cost and installation of this equipment, which operates on a simple computer dedicated to the control system, is the easy part.

The hard part is the commitment and education required to apply all of the available programs to enhancing the throughput or yield of salable high quality castings. Every die must be monitored with hard copy confirmation that is then input into an X bar and R chart where trends may be studied and operating decisions made that will directly affect the throughput.

In the case of a difficult die with a small operating window, process monitoring is used as a diagnostic tool to identify and solve problems in a quantitative style. Many times Design of Experiments (DOE), the Taguchi technology, imports the monitoring data to identify the variables that are the root cause and must be brought into control.

It is just not as simple as hooking up a computer and generating a bunch of charts that end up in a file cabinet. Monitoring has to be viewed as part of the decision making process. Too often the engineers and technicians that are qualified for this work have experience dealing with emergencies rather than strategically preventing the emergencies from ever occurring.

The PQ Squared Concept dates back to the 1970s when the Commonwealth Scientific Industrial Research Organization (CSIRO) developed a method to graphically describe the power characteristics of the shot end of the high pressure die casting machine [ADCA, Die Casting Bulletin (Jan/Feb 1995)]. The name is derived from the relationship of the pressure on the metal (P) to the quantity or flow rate (Q).

Since the flow rate is a function of the velocity of the shot plunger, the concept is sometimes referred to as the PV Squared Concept as well. It is widely used by researchers, machine manufacturers, and working die casters because the physical principles are the mathematical basis for the supply of liquid casting alloy to the die cavities.

A broad review of the concept should be helpful. It is based upon Bernoulli's equation that is defined for this purpose as $V = k2P/d$, where $V = velocity$; $P = pressure$; $d = liquid\ density\ of\ the\ casting\ alloy;\ k = discharge\ coefficient\ or\ efficiency.$

A discharge coefficient is not as critical with a cold chamber process since the shot sleeve is straight. Therefore, a value above 97% or > 0.97 can be used. This is not the case with the hot chamber process, however, where the losses in the goose neck caused by abrupt directional changes in the flow path seriously effect the efficiency. The nozzle also adds to the problem so a value of k should be in the range of 50–60% or < 0.60 to be realistic.

When a particular shot sleeve or goose neck is combined with the casting machine shot system, the pumping mechanism is complete and a value for quantity (Q) can be calculated by the hydraulic formula $Q = AV$. Then the equation becomes $Q = Ak2P/d$.

It is a good practice to establish the maximum pressure and dry shot velocity for a particular machine by measuring the dry shot capability within the pressure range recommended by the machine manufacturer. The mathematical information is included here to explain the theory and logic behind the concept.

The equations illustrate the squared relationship between pressure and velocity with respect to fluid flow. Thus, parameters plotted upon a diagram where the pressure scale is linear and the Q or V scale follows the squared rule will be represented by straight lines.

The concept is explained by the following diagrams and discussions (Von Tachach, 1996).

The pressure scale in Fig. 9 is linear and drawn on the Y axis. The Q scale is quadratic on the X axis where divisions

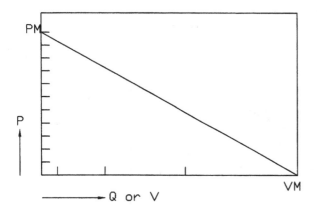

Figure 9

represent 1, 2, 3, and 4. *PM* is the maximum pressure and *VM* is the maximum dry shot piston and plunger velocity which is determined by measurement.

Line *PM–VM* in Fig. 9 defines the maximum machine power or capability to pump liquid casting alloy. The area outside the triangle as well as the origin cannot be used.

Figure 10 adds the restriction that different gate areas apply to the system. Lines g_1 through g_5 are called gate lines where g_1 describes the smallest and g_5 the largest gate area studied. It can be easily seen that flow rate increases with gate area when all other variables remain constant.

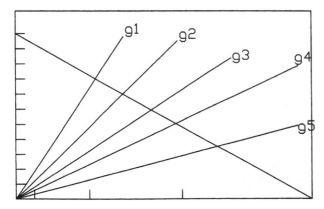

Figure 10

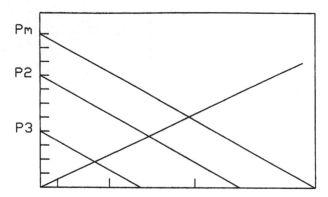

Figure 11

Figure 11 shows what happens when hydraulic pressure is changed. Increased pressure also increases flow proportionally to the change in pressure. The three pressure lines are parallel where P_1 is the greatest and P_3 is the lowest pressure. These lines are connected by a specific gate restriction and it is clear that a reduction in pressure also reduces the flow rate and vice versa (Fig. 9).

Figure 12 offers a pqsq. diagram that describes what happens to the flow rate (Q) when the setting on the shot velocity control valve is changed to revise the flow of hydraulic

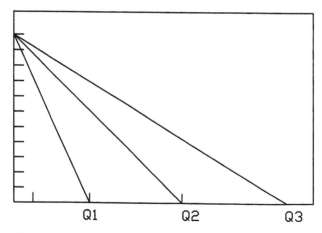

Figure 12

fluid into the shot cylinder. This also changes the corresponding flow rate of the casting alloy in that Q_1 is less than Q_3.

The pressure does not change but when the shot piston reaches its maximum displacement, the pressure will rise to the maximum.

The liquid density of the super heated casting alloy has a profound influence upon the flow rate, which is described in Fig. 13. Sometimes, this is not fully appreciated, but all die casters are aware of fluid flow differences in casting alloys. The two lines m_1 and m_2 portray casting alloys of different liquid density where m_1 has a higher density than m_2. The flow rate from the alloy with higher density is lower even though the available pressure (P) does not change. Of course, the gate area of each is identical as well.

Figure 14 shows how flow rate is affected by change in diameter (area) of the shot sleeve (cold chamber or goose neck). This is diagramed with $ss1$, $ss2$, and $ss3$. When the area of the shot sleeve is increased, the flow rate and the pressure requirement is decreased. The gate line shown, of course, utilizes the same gate area was used for all three.

Figure 15 illustrates the difference between individual die casting machines which sometimes vary even though they are of the same vintage, locking force, and make. A unique characteristic can be seen in that Mch_1 displays a low flow rate with a high pressure and Mch_3 produces just the opposite

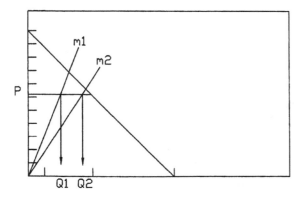

Figure 13

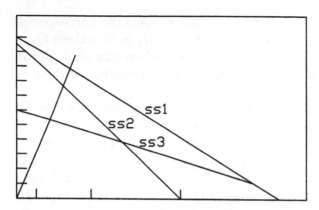

Figure 14

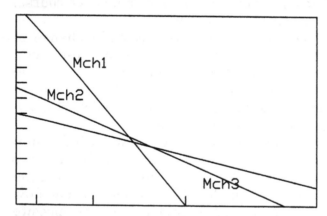

Figure 15

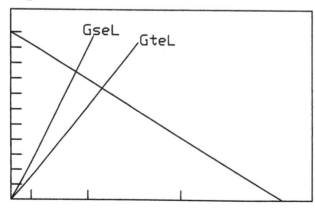

Figure 16

with a high flow rate at low pressure. This is the reason for performing a dry shot measurement to characterize the shot performance of all machines.

The pqsq diagram in Fig. 16 describes a hot chamber machine—a very different situation because the fluid flow losses are huge. The cold chamber process discussed up to this point is considerably more efficient with negligible losses that are many times ignored. Line *GseL* graphs the losses in the goose neck and the restriction of the nozzle. *GteL* is the gate line that shows the restriction of the gate. Note that more than half of the machine power is required merely to pump the liquid metal to the nozzle.

Even though the hot chamber process is hydraulically inefficient, it is the most thermally and economically efficient of the two processes. This explains its popularity for all alloys except aluminum that acts as a solvent whenever ferrous materials like cast iron goose necks are immersed into a bath of this liquid metal.

Figure 17 shows the relationship of different gate areas to the machine power line. Line g_1 describes the smallest gate area and line g_3 the largest. It can be seen that the larger gate area generates the greatest flow rate with the lowest pressure applied to the metal.

One application of the diagram is described in Fig. 18. *CF* is the pressure (p) and flow rate (q) during cavity fill and P is the pressure after the plunger movement has stopped at the

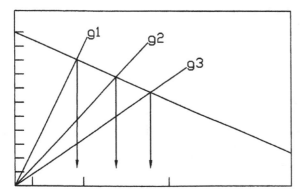

Figure 17

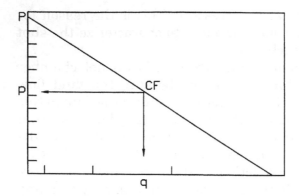

Figure 18

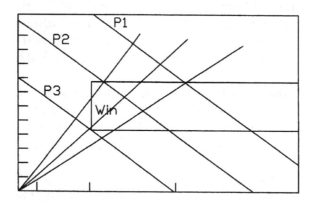

Figure 19

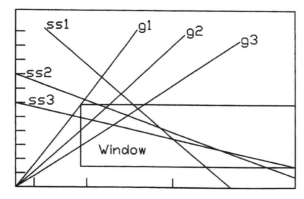

Figure 20

end of cavity fill. This graph is useful after the range of cavity fill time, gate speed, and gate area has been calculated.

Figure 19 starts to define the operating window (*Win*) by describing the acceptable pressure range of the shot system of the die casting machine within the three pressure lines. P_1 is too high and P_3 is too low. The two horizontal lines limit the acceptable extremes of gate areas.

Figure 20 a diagram that is used to determine the area of the shot sleeve. Gate area g_1 is on the edge of the window but the other two choices are well within it. The three sleeve choices are $ss1$, $ss2$, and $ss3$ with $ss2$ being the best choice since it and the two gate options of choice are centered in the window.

Figure 21 establishes the range of acceptable gate speeds, which are pressure dependent and limited by the degree of difficulty in casting the product, quality requirements, and die life when aluminum is the casting alloy.

Figure 22 deals with cavity fill time, which is a function of the flow rate. The maximum fill time is defined by the ratio of surface area to volume, plus distance that the liquid casting alloy must travel within the net shape of the cavity. The thermal behavior of the metal also is significant to this scenario.

Thus, the horizontal dimension of the operating window is determined by the acceptable cavity fill time range. Figure 23 defines perfect limits.

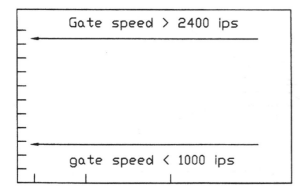

Figure 21

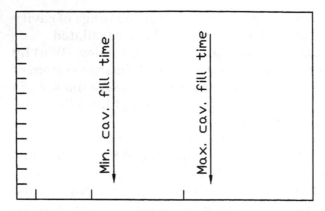

Figure 22

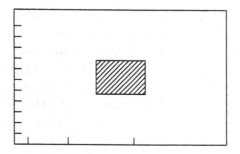

Figure 23

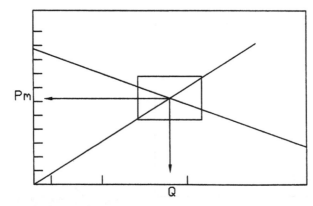

Figure 24

The previous two diagrams define the limits of the operating window.

Figure 24 pictures an ideal combination of die casting machine shot system, shot sleeve, and the metal feed system in the die that centers in the operating window. This centering allows the die to continue to produce castings of acceptable quality even though process variables may deviate from the ideal settings.

The *PQ* Squared diagram illustrated in Fig. 25, demonstrates the versatility of the concept, in that the operating window (*WIN*) is defined by a range of reasonable plunger velocities that are within the dry shot capability of the casting machine shot end. These velocities are constant on the diagram. The restriction lines represent a range of gate areas that meet acceptable thickness criteria to accommodate the casting alloy and trim requirements. The machine line is the third limit that identifies the operating window.

Pressure as indicated on the *Y* axis of the diagrams refers to hydraulic pressure on the shot piston or that pressure as it is applied to the metal. The hydraulic pressure is defined by the area of the shot cylinder; the pressure on the metal is calculated over the area of the shot sleeve or goose neck.

The theory of the PQ Squared concept is verbally and graphically explained above, but to complete this discourse, it is important that the serious die caster be able to make the necessary calculations and to construct a PQ Square

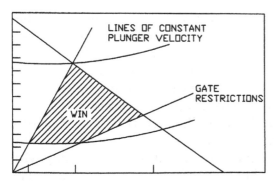

Figure 25

diagram. This information and the appropriate equations are outlined on the following pages.

To construct a PQ Squared diagram, the following information is necessary:

- Hydraulic pressure on the shot cylinder (accumulator pressure × area of cylinder).
- Net cross-sectional area of shot piston (Piston area-ram area).
- Maximum dry shot plunger velocity.

Parameters required:

- Hydraulic pressure range of machine shot system.
- Flow characteristics of shot control valve (% open compared to % of maximum flow).
- Locking force of machine.

Calculations for diagram input data are:

- Pressure on metal = Net shot piston area × Net hydraulic pressure/Area of shot sleeve
- Maximum calculated flow rate = Maximum dry shot velocity × Largest shot sleeve area
- Injection force = Pressure on metal × Area of shot sleeve
- Actual flow rate (quantity) = Plunger velocity × Area of shot sleeve ($Q = AV$)

Note that the shot sleeve refers to goose neck area in the hot chamber process. Hot chamber shot systems are inefficient and the above calculations must be multiplied by the discharge coefficient which is considerably less than 1 (approximately 60%).

The Scale of the *PQ* Squared diagram is quadratic and can be constructed by horizontal lines to represent the pressure that is linear on the *Y* scale. The quantity (*Q*) or flow rate must be calibrated on the *X* scale by squaring each increment. Starting with 1 in.^{3}/sec squared = 1 and then continuing with 2 squared = 4, etc. The blank scale in Fig. 26 can be set up as a basis for plotting the data.

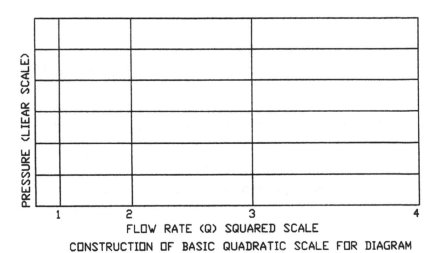

FLOW RATE (Q) SQUARED SCALE

CONSTRUCTION OF BASIC QUADRATIC SCALE FOR DIAGRAM

Figure 26

To calculate the maximum metal pressure:

Select the smallest shot sleeve that can be used on the machine since this will maximize metal pressure and minimize back scrap.

As an example, a 3 in. diameter is chosen for a machine with a shot cylinder diameter of 6 in., 3 in. diameter ram, and a maximum pressure limit of 1800 psi.

We will assume a discharge coefficient of 98%.
Then:

Piston area $= \pi \times 36/4 = 28.26$ in.^{2}

Ram area $= \pi \times 9/4 = -7.07$ in.^{2}

Area of shot sleeve $= \pi \times 9/4 = 7.07$ in.^{2}

Net piston area $= 28.26 - 7.07 = 21.19$ in.^{2}

$P_m = 1800 \times 21.19 \times 98/7.07 = 5287$ psi

To calculate the maximum flow rate:

Select the largest shot sleeve that can be used on the machine because this will maximize the flow rate. For this example, a 4 in. diameter is picked and the maximum dry shot plunger velocity is 240 ips.

Then,

$$Q = \pi \times 16/4(240) = 3014 \, \text{in.}^3/\text{sec}$$

The machine power line can now be constructed as illustrated in Fig. 27.

What if a 3 in. diameter shot sleeve and plunger were used?
Then,

$$Q = 240 \times 7.07 = 1697 \, \text{in.}^3/\text{sec}$$

This line can be plotted on the diagram in Fig. 26.
What if the shot pressure were redeuced to 1600 psi?
Then,

$$P_m = 1600 \times 21.19 \times .987/7.07 = 4733 \, \text{psi}$$

This line is plotted on the diagram offered in Fig. 27.

To summarize machine related parameters:

- Given two different size shot sleeves, the smaller one will place more pressure on the casting alloy if the hydraulic pressure is the same.

- At the same hydraulic pressure, the larger shot sleeve will generate a higher flow rate.

- Identical shot sleeve or goose neck diameters at different hydraulic pressures will be represented by parallel lines on the *PQ* Squared diagram.

- Machine power curves at different hydraulic pressures will be described by parallel lines on the diagram.

- The shot sleeve area and the pressure on the casting alloy have an inverse relationship.

- The shot sleeve area and the flow rate have a direct relationship.

- Flow rate influences cavity fill time.

- Both pressure and flow rate affect gate speed.

Process monitoring of shot system repeatability is very common in the die casting industry and several good

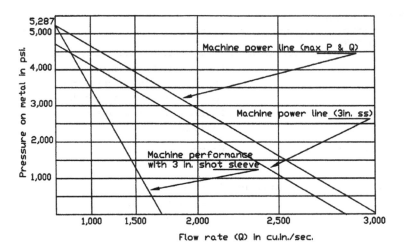

Figure 27

computerized monitors are commercially available. Almost all variables can be measured with a wide array of sensors that feed data into the computer hardware. There it is processed by appropriate software to accurately measure variables when configured into final data.

For purposes of simplification here, the critical variables that define process control and what they affect are outlined in Table 2. Temperatures of the casting alloy as it passes through the various phases from ingot to net shape can indeed be successfully managed.

Table 2   Product Variables and Controls

Critical variable	Control
To control volume of metal supply	Biscuit thickness (cold chamber only)
To control chamber (cold chamber only) and runner fill	Slow shot plunger velocity
To control turbulence and cavity fill time	Fast shot plunger velocity Gate speed
To control metal temperature at the end of cavity fill	Cavity fill time

(Continued)

Table 2   (Continued)

Critical variable	Control
	Die surface temperature at end of cavity fill, at a specified strategic location in each die half Holding furnace temperature Cycle time
To control dimensional stability	Ejection temperature of cast shape at last detail to solidify
To control internal casting integrity	Accumulator pressure Pressure rise time Intensification pressure

Die surface temperature at the end of cavity fill must be measured by thermocouples embedded in the die steels. Of course, these thermocouples do not see the interface between the super heated liquid casting alloy and the surface of the die steel because they are not designed to withstand the extreme heat and abrasion. Therefore, these data are useful for bench marking purposes only. Any deviation in temperature signals an unacceptable event.

A step beyond monitoring is to program the data for each casting cycle into the rest of the die casting cell. If the robot or extractor is set up to place an extracted shot into the main production flow only when all variables are within established analytical parameters, only good and salable product can enter the throughput.

Since there is a safety factor calculated into each parameter, it is possible that a good portion of the side-tracked product may be acceptable and salable. Extreme care must be taken with human inspection to ensure that throughput is not contaminated.

Sophisticated process control described here is not the norm in the die casting industry. It certainly is not easy and requires expert professional, computer-literate, and skilled electronic technicians. There is no question that all equipment in the die casting manufacturing cell has to be in exceptional operating condition. It need not necessarily be new, but must be continuously maintained for repeatability and exceptional operating condition.

Minimum process control is described above. Without all of it in place, there is no statistical process control and the quality requirements of automotive die casting users become merely acronyms and paper work to keep die casting suppliers on the bidders list. APQP is a good example. It stands for Advanced Product Quality Planning. How can quality really be planned without quantified process control?

An even better method for process control is more complicated and expensive, but embraces the die casting user's perspective in a more modern manner.

Real time closed-loop control of the shot end has been commercially available for quite some time to remove all human decisions from the dynamic dimensions of the shot end (Hedenhag, 1989). The variable factors can be divided into three categories—static, manual, and dynamic.

Static factors include the condition of the die casting machine, design and condition of the die, and accumulator and intensification pressures. Establishing metal temperature and viscosity of the hydraulic fluid, and setting of valves, limit switches, and timers are manual factors that are subject to human influence. More difficult to control are the dynamic factors of volume of metal in the sleeve or goose neck, plunger drag, die temperature, and the vacuum pressure profile.

The metal temperature and other factors will inevitably vary from shot to shot, regardless of how hard we try to keep them constant. The greatest variations fall into the dynamic category. Variations in plunger velocity and pressure are substantial and deviate from an ideal shot trace from cycle to cycle.

Shot control compensates for dynamic variations from the ideal by correcting deviant parameters during the shot! Given the average cavity fill time is between 20 and 80 msec, it has been determined that correction must occur in approximately one-tenth of that time or between 2 and 8 ms, to effect the desired repeatability.

For this purpose, total response time is defined as the time elapsed from the instant the sensor detects a deviation to the instant the plunger starts to react. The response time for the hydraulic system, including valve and shot cylinder delays, puts heroic demands upon the electric circuitry, which must respond within a few milliseconds.

An adaptive control system that corrects shot parameters based upon the previous cycle is of limited value because it is unable to adjust for the large and inconsistent variables that take place from shot to shot and within each cycle. The real-time closed-loop system controls actual plunger velocity by the hydraulic pressure in the shot cylinder of the casting machine. The pressure continuously changes so that the plunger velocity follows the master shot profile.

ISO, QS, Six Sigma, Lean Manufacturing, and JIT (discussed briefly in Chapter 10) are only documentation without controls to restrict human decision. This chapter was intended to explain methods to close the gap between documentation and predictable quality and productivity. There is a test to prove the fallibility of human accuracy. If 10 white balls are included in 990 black balls, it is impossible for the human mind and behavior to separate them properly on the first try. Yes, sorting good ones from bad ones requires too much manufacturing energy and is not accurate enough.

9

A Thermal Process

Die casting is basically a thermal process even though the preceding chapters have discussed a variety of mechanical and hydraulic procedures. After considerable technical involvement in many different die casting plants all over the world, this writer has been privileged to observe that only minimal attention is directed toward the thermal dimension of the process. The focus in this chapter is upon the sometimes fine line between what works and what does not work.

As a result, only a few dies operate at their maximum capacity to distribute heat energy evenly and then remove it efficiently. In other words, the yield of acceptable castings produced by the typical die casting die is nearer the 80 than the 100 percentile of efficiency.

At this writing, the hourly cost to operate a die casting machine (150–2500 ton) is between \$200.00 and \$1,000.00. The 20% gap between actual and possible translates into tremendous economic potential!

This chapter discusses the accepted rules to follow to maximize the life of the die and to produce quality castings. One die casting firm in North America breaks some of the

rules by excessive external cooling (high-velocity application
of cold water at each cycle) by balancing die replacement cost
against production cost economy on high volume parts. This
pressing of the operating envelope must be paying big dividends.

Great heat energy is required to superheat the casting
alloy into the liquid state so that the viscosity approximates
that of water. It is important that each alloy used in the die
casting process behaves as a hydraulic fluid during the cavity
fill phase. This heat energy is quantified in either British
thermal units (btu) or joules (j) when the metric system is
employed. English units are used for the purpose of this text.

For perspective, a btu is defined as the heat energy
required to raise the temperature of 1 pound of water by
1°F. Thus, the specific heat of water is 1 btu. With this analogy in mind, analytical calculations can be made to define
the heat energy that impacts the die surface during the filling
of the cavity to form the net shape of the casting. Such calculations are critical to die casting production, but it is also
important to monitor the heat energy at different points in
the process such as melting, transfer, ladling (cold chamber)
or pumping (hot chamber), and injection. Energy can be
calculated but not measured, so temperature is the unit of
measure.

The dynamic nature of temperature/heat energy of typical aluminum alloy die casting cycles is demonstrated in
Fig. 1. It is apparent where high temperatures are necessary
to maintain the liquid state of the casting alloy and where

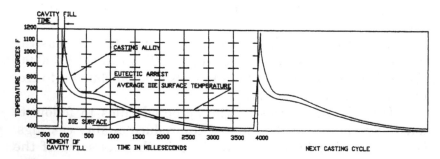

Figure 1

rapidly decreasing temperatures convert the casting alloy back to the solid state prior to ejection.

Notice how the large gradient between metal temperature and die surface temperature during cavity fill closes dramatically during solidification that occurs during the dwell time of the casting cycle. The casting alloy is superheated in the breakdown (melting) and holding furnaces prior to delivering it to the injection chamber. At this extremely elevated temperature, the viscosity reduces to about that of water at room temperature so that the alloy will behave like a hydraulic fluid while the near net shape is formed during the filling of the cavity.

Superheat varies with each base metal and is sometimes referred to as the heat load. The base metal is the predominant metal in an alloy system. At the end of cavity fill, enough of the superheat must be rapidly removed to solidify the alloy sufficiently to survive the forces of ejection from the die. There is a definite schedule to this freezing process that is dictated by the casting material.

Pure metals that are used in alloys for die casting have different cooling patterns as described in Fig. 2. Heat is lost as the length of time away from the energy source increases. However, temperature does not change even though heat continues to be lost when the metal changes from the liquid to the solid state as noted by the flat segment of each curve. This is called the eutectic or lowest melting point of the base metal.

The predominant metal in an alloy system is called the base metal. It is customary to express this metal first in naming each system. Conversely, when casting alloys are superheated, this flat portion of the curve requires the latent heat of fusion to convert the alloy from solid to liquid. This is covered in more detail later in the chapter.

Die casting alloys behave a little differently from pure metals because of the alloying with metals other than the base metal necessary to satisfy the vicissitudes of the die casting process. Fig. 3 is based upon two of the most popular aluminum alloys and shows that the flat element of the curve is not totally flat during solidification.

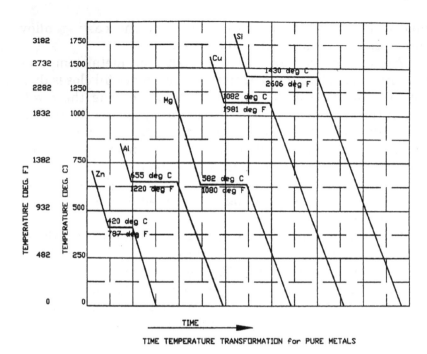

TIME TEMPERATURE TRANSFORMATION for PURE METALS

Figure 2

Aluminum alloys were chosen for Fig. 3 because they constitute at least 70% of the casting material consumed. As an aside, it should be noted that the freezing range of the more common 383 alloys is considerably greater than the 413 alloys because the latter is designed to contain silicon right at the eutectic point of 12.6%. This gives it a premium cost and relatively tight solidification range. This also limits the opportunity for volumetric reduction that significantly reduces the possibility for shrinkage porosity, one of the major defects generated by the high pressure die casting process.

The role of the casting die as a heat exchanger is obvious from the discussion thus far because raising the casting alloy to the injection temperature is not a part of the die casting process! It is the pattern and speed at which the superheat energy is removed between cavity fill and ejection that are critical. *It is all about heat removal.*

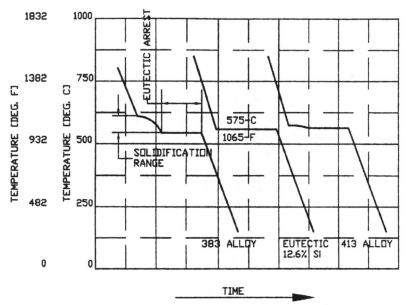

TIME TEMPERATURE TRANSFORMATION for ALUMINUM ALLOYS

Figure 3

The desired dynamics must be controlled by a thermal strategy that is designed into the die casting die and operating process. It is called a thermal system here and not a cooling system because, while it is usually necessary to remove heat quickly to achieve rapid solidification, sometimes it is also important to retain heat longer in certain regions of the casting to achieve challenging casting quality objectives.

The design of the thermal system is important because it affects (CSIRO, 1991):

- Casting quality
- Casting production rate (economics)
- Die life (economics)

In practice, several thermal circuits are drilled through the die steels that carry either water or oil that acts as the heat transfer medium. Water always removes heat, but high temperature oil can both heat and cool the die component.

Oil is not as efficient as water for cooling, which requires a larger and longer circuit.

Thus, the die acts as a heat exchanger that helps to convert the superheated casting alloy from the liquid to the solid state in a very short period of time. The conversion of metal in the liquid state to a net shape in the solid state in milliseconds requires knowledge of the difference between temperature and heat.

It is difficult because, at the levels useful to die casting, heat energy cannot be seen, has no odor, and cannot be touched. This may be the reason why many die casting operations take the thermal aspects as they come, without monitoring or control.

As noted, the design of the thermal system in the die casting die profoundly affects the quality of the castings, the production rate, and the life of the die casting die ... all of the elements that determine the profitability of the business.

Heat is energy and therefore has quantity that can be measured, controlled, and monitored. Temperature is the heat equivalent of pressure, and when measured, the amount of heat can be calculated, using the size and heat absorbing properties of the material. Thus, heat in a block of steel, such as a die casting die, will escape rapidly if the temperature is high.

Heat is transferred by the mechanisms of conduction, convection, and radiation, which are described here in some detail—it is important to understand the principles of heat transfer during the dwell phase of the die casting process.

Conduction, the greatest factor, occurs when heat moves from a higher temperature to a lower temperature within a single die component. It depends upon several things:

- Material of the conductor
- Mass of the conductor
- Distance that heat must travel
- Temperature gradient between die/casting alloy interface and point of interest (i.e., water line)

The conductor is the die component that moves the heat; the air space between the die component and the retainer is just the opposite, an insulator. H-13 die steel is not the best conductor—it is the material of choice because of its fracture toughness rather than for its thermal conductivity. Copper is an excellent conductor and is therefore used in North America as the material for plunger tips since the biscuit is usually the most massive detail in the shot. The thermal conductivity of H-13 die steel is 1.93 btu/in./°F and beryllium copper has a conductivity of 4.81 btu/in./°F.

The mass of the die component affects heat transfer by providing the space for conduction. The larger mass provides less resistance and therefore conducts heat more easily than a smaller area. Thus, a good case can be made for larger die components.

Distance inversely affects heat flow. Less heat will flow over a longer distance than a shorter one. Therefore, cooling channels located farther from the die/alloy interface will remove heat slower than if they were closer.

The temperature difference between the source and the heat sink is the force that drives the heat. Greater differences increase heat transfer. Heat will not move if there is no temperature difference; we call this a balanced condition.

Convection is another important mechanism for heat transfer and takes place when cold fluid passes through a hot die component. Natural convection happens because the die retainer surfaces that are exposed to ambient air are hotter than the air. Heat moves from hot to cold. The same convection would be forced if a fan were directed at the hot die surfaces. Internal cooling occurs within the die casting die because the colder medium flowing through the thermal channels removes the heat conducted to the channel location by force.

The variables that characterize the convection mechanism are:

- Convection film coefficient
- Contact area for heat transfer
- Temperature difference between hot die surfaces and the cooling mediums

The convection film coefficient is complex in that it is a function of several variables. It makes a difference whether the convection is forced or not. In the case of internal cooling in a die casting die, it is forced. The medium used is a variable because the usual fluids utilized, oil and water, behave differently. The velocity of the medium flowing through a channel has a profound effect upon the convection rate. Believe it or not, the same flow rate through smaller diameter channels offers more convection than through larger lines since the velocity is faster. This is not a good reason to opt for smaller channels, though, since convection can easily be increased merely by a higher flow rate.

Table 1 *quantifies convection film coefficients* under several different conditions and may be useful to the serious student of heat removal from a die casting die.

Table 1   Convection Film Coefficients

Cooling medium	Channel diameter (in.)	Flow rate (gpm)	Convection film coefficient (btu/hr/sq.in./°F)
Water	7/16	1.0	3.5
Oil		1.0	1.4
Water		2.0	6.1
Oil		2.0	2.4
Water	9/16	1.0	2.4
Oil		1.0	0.96
Water		2.0	4.2
Oil		2.0	1.68
Water		3.0	5.7
Oil		3.0	2.28

Notice the dramatic difference between the two most popular mediums. Heat is removed more gently by oil so the thermal shock to the die steels is much less than water and a much longer die life can be expected. The usual logic is to cool massive elements of the shot like runners, sprues, biscuits, and more massive cavity and core details with water. Oil is the choice for cavities and cores where the channel surface is close to the metal/core interface.

Radiation causes the die to lose some heat, but only from surfaces that are exposed to the ambient air. Air is considered a fluid and when it moves across the surface of the die, the fluid (air) absorbs the heat from the hot surface and carries it away into the atmosphere. Thus, the space around an operating die casting machine feels hot.

In die casting, we usually think of radiation when superheated liquid metal is held or moved and heat is lost when it radiates from the surface into the atmosphere.

Superheated liquid metal is the main source of the heat input, which is cyclical. Therefore, it is more useful to consider the rate of heat input, heat loss, and heat absorbtion. If the cooling capacity is not sufficient, the die temperature will increase. Conversely, if the loss to ambient air is too great, the die temperature will drop.

Heat input into the die casting die from the liquid casting alloy can be calculated but depends upon the variables listed below:

- Specific heat
- Latent heat
- Mass of the casting
- Injection and ejection temperature of the casting
- Production rate

Specific heat of a material is the amount of heat required per unit of mass of the material to raise its temperature by one unit. Table 2 illustrates the specific heat for typical die casting alloys and die steels. Please note that there is not too much difference in the values of casting alloys when compared on the basis of volume.

Table 2

	Specific heat	
Material	Weight basis	Volume basis
Water	1.0 btu/lb/°F	0.036 btu/cu.in.°F
Zinc	0.1	0.024
Aluminum	0.26	0.025
Magnesium	0.34	0.016
Brass	0.10	0.027
Die steel	0.11	0.100

One British thermal unit (btu) is the amount of heat required to raise the temperature of 1 pound of water by 1 degree Fahrenheit (°F). Thus, the specific heat of water equals 1 btu/lb/°F.

In die casting, the liquid specific heat rather than the solid specific heat is of more interest since it is more beneficial to cast an alloy when it is in the liquid state during cavity fill. The alloy experiences a slushy state in the range between the liquid and solid specific heats.

Latent heat of fusion is the amount of heat required per unit mass of material to convert it from solid to liquid. Latent heats for some casting alloys are listed in Table 3.

Table 3

Material	Latent heat	
	By weight	By volume
Zinc	43.0 btu/lb	10.6 btu/cu.in.
Aluminum	169.0	22.5
Magnesium	157.9	10.6

The thermal behavior of popular casting alloys is illustrated in Fig. 4. The heat load of aluminum alloys as compared to zinc and magnesium is described by the significant difference. Longer casting cycles can be expected with aluminum because there is so much more heat to remove during each cycle. More careful design of the thermal system is called for to optimize productivity.

This writer has had the opportunity to quantitatively examine casting productivity over a large cross-section of the worldwide die casting industry after observing casting operations at hundreds of plants, I can state that no die casting firm, save one, ever achieves maximum shots per hour because not enough attention is paid to calculating the real time effect of this behavior.

The heat in the casting alloy can be easily calculated; it is the heat load that is placed upon the die each and every casting cycle. To do this, it is necessary to calculate the heat energy required to superheat the melt up to the desired

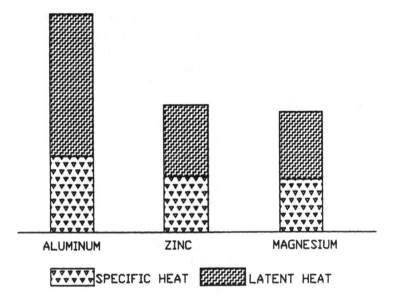

Figure 4

holding temperature. Visualize this process as occurring in
two steps. First, the metal temperature must be raised from
ambient room temperature up to the solidus temperature.
Then, the latent heat of fusion forces the metal through the
eutectic arrest (see Fig. 3) to convert the metal from the liquid
to the solid state. The sum of these two stages brings the
liquid metal to the desired superheat.

The quantitative formulae that calculate the heat input
are stated as follows:

$$Q_s = V \times H_s(T_s - T_a)$$
$$Q_f = V \times H_s(T_1 - T_s) + H_f$$
$$Q_{sh} = V \times H_s(T_{sh} + T_1)$$
$$Q = Q_s + Q_f + Q_{sh}$$

where

$Q =$ Total heat required to super heat casting alloy from
ambient room temperature to holding furnace
temperature

Q_s = Heat required to increase casting alloy from ambient room temperature to the solidus temperature

Q_f = Heat required to increase casting alloy from the solidus to liquidus temperature

Q_{sh} = Heat required to super heat casting alloy to holding furnace temperature

V = Volume of casting alloy in shot

H_s = Specific heat of alloy,

H_f = Latent heat of fusion of alloy.

To put some numbers to these formulae, a hypothetical shot with 100 cu.in. volume (9 lbs) of 380 aluminum alloy is illustrated here to calculate the heat energy necessary to raise an ingot from 70°F ambient room temperature to a superheat 1250°F in the holding furnace.

$$Q_s = 100 \, \text{cu.in.} \times 0.025 \, \text{btu/cu.in./°F}(968°F - 70°F)$$
$$= 2245 \, \text{btu}$$
$$Q_f = 100 \, \text{cu.in.} \times 0.025 \, \text{btu/cu.in./°F}(1094°F - 968°F)$$
$$+ \, 100 \times 22.5 \, \text{btu/cu.in.} = 2565 \, \text{btu}$$
$$Q_{sh} = 100 \, \text{cu.in.} \times 0.025 \, \text{btu/cu.in./°F}(1250°F - 1094°F)$$
$$= 390 \, \text{btu}$$
$$Q = 2245 \, \text{btu} + 2565 \, \text{btu} + 390 \, \text{btu} = 5200 \, \text{btu}$$

In this scenario, since each shot contains 100 cu.in. of 380 aluminum, 2975.5 btu are poured into the cold chamber every casting cycle. To balance the thermal equation, 2975.5 btu of heat energy must be removed during each cycle.

Superheat loss between the holding furnace and the gate is explained in Chapter 5 with some comprehensive nomographs that reveal approximately 50°F loss. The heat energy can be calculated for this example as follows:

$$Q_{loss} = 100 \, \text{cu.in.} \times 0.025 \, \text{btu/cu.in./°F}(1200°F - 1250°F)$$
$$= -125 \, \text{btu}$$

The heat energy that must be removed is calculated with the same formulae with negative numbers that describe the losses. Therefore, using the same example with a target ejection

temperature of the casting of 500°F, these calculations establish the amount of heat removal required each casting cycle.
The ejection temperature of the casting is important here
because the heat loading ends when the shot leaves the die.

The first calculation removes the heat between the
1200°F injection temperature and the liquidus temperature
at the eutectic and looks like this:

$$Q_{sh} = 100\,\text{cu.in.} \times 0.025\,\text{btu/cu.in.}/°\text{F}(1094°\text{F} - 1200°\text{F})$$
$$= -265\,\text{btu}$$
$$Q_f = 100\,\text{cu.in.} \times 0.025\,\text{btu/cu.in.}/°\text{F}(968°\text{F} - 1094°\text{F})$$
$$- 100 \times 22.5\,\text{btu/cu.in.} = -2565\,\text{btu}$$
$$Q_s = 100\,\text{cu.in.} \times 0.025\,\text{btu/cu.in.}/°\text{F}(500°\text{F} - 968°\text{F})$$
$$= -1170\,\text{btu}$$
$$Q = -265\,\text{btu} - 2565\,\text{btu} - 1170\,\text{btu} = 4000\,\text{btu}$$

The difference between 1200 btu of the heat load of 5200 and
4000 btu is ejected away from the die with the shot that is 500°F.
The ejection temperature should *never* be set arbitrarily since it
has a profound effect upon the productivity of the die. Just put in
an ejection temperature of 300°F, and see what happens!

Normally, the casting shot is quenched to bring it closer
to room temperature prior to trimming. In this case the
energy that the quench must remove can be calculated by
one more formula: $Q_q = 100\,\text{cu.in.} \times 0.025\,\text{btu/cu.in.}/°\text{F}$
$(70°\text{F} - 500°\text{F}) = 1075\,\text{btu.}$

Productivity in casting cycles per hour is a function of the
heat load calculations and the capacity of the die casting die
to remove heat, which is based upon the heat transfer
mechanisms. For example, heat removed by conduction can
be calculated with this formula:

$$Q = C \times A(T_i - T_o)/D$$

where Q is the heat transferred per hour, C the coefficient
of thermal conductivity, A the area of die component, T_i the
temperature at source, T_o the temperature at sink, and D
the distance.

Removal of heat load from the die casting die cannot be taken lightly as it affects both the quality of castings produced and casting cycle time, which deals with cost. Quality is supported by consistent maintenance of die surface temperature so that liquidity of the casting alloy is ensured during cavity fill. Ideal targeted die surface temperatures are discussed elsewhere in this book, but it is critical that analytical logic be incorporated in the design of the thermal system for the die casting die. Since cavity fill time represents only a very small portion of the total cycle time, it is not important to the cost of operation. However, heat removal is critical to optimum cycle speed (shots per hour). Therefore, the important aspects of die design are now offered.

Heat is conducted through the die components in a straight line until some other force changes the direction of heat movement. The route followed is called a heat path. The thermal paths must be understood so that they can be identified if the thermal system is to be properly designed. This is too involved a subject to be covered in depth in this text, but basically, thermal paths converge away from the casting and into a core in a die casting die and diverge away from the cavity. This concept is presented in Fig. 5.

From this simple illustration, it can be seen that heat paths converge into the core and diverge away from the cavity. According to Boyle's law, heat increases proportional to its mass and decreases as the mass reduces. Thus, the core can be expected to operate at a higher temperature than the cavity and will require considerably more cooling.

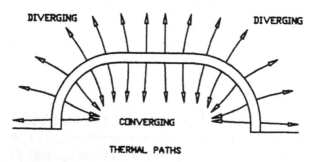

Figure 5

It is critical to control the die surface temperature so that it is uniform by providing a thermal system for cooling and/or heating (on rare occasions external heating is used). The system has to have sufficient capacity and be located correctly to ensure that the die temperature can be controlled at the appropriate level. This is not a simple determination of size and length of thermal circuits in the die block.

The position must be determined correctly to deal with the thermal paths as naturally as possible. The goal is to maintain the die surface temperature at the casting/die interface that will produce quality castings.

If the cooling channel is located too far from a particular die detail, the surface temperature becomes too hot and heat-related defects like blisters or solder will be experienced. The other extreme is to locate the cooling circuit too close to the die detail, which causes the die surface temperature to run too low and can lead to cold shut or lamination defects.

This must all be strategically determined before the die steels are machined and hardened since it is not practical to change the thermal system after they are finished. In practice, deficiencies are sometimes discovered after castings are produced, which forces the alternative of excess external die spray to remove the additional heat. This works, but slows down the production rate and seriously reduces die life.

A simple explanation of the heat flow process in die casting dies may be useful in understanding how heat is exchanged between the casting alloy and the die, and then is conducted away through the cooling medium.

The superheated casting alloy is the source of heat energy that is conducted by the die steel to the surface of the cooling channel (circumference). The amount of heat transferred is a function of the difference between the temperatures of the die steel minus the temperature of the cooling medium (water or oil) multiplied by the heat transfer coefficient of the medium.

A similar condition is experienced between different die components. The amount of heat transferred is the temperature difference between the two separate steels divided by the distance to the next interface and multiplied by the heat transfer coefficient of the die material.

The heat transfer coefficients at the interface between the casting and the die surface vary with the casting temperature, die surface condition, and the solidification process. Therefore, heat flow between casting and die is difficult and time consuming to calculate, so in most die casting die designs this important calculation is not made. Fortunately, there are computer aided engineering programs available to quickly and accurately perform this mathematical task. It is more complex than the early manual methods, which are now outdated.

Caution must be observed when calculating the heat exchange between two separate die components because they seldom fit tightly together. The air gap, usually in the range of 0.002 in., acts as an insulator that retards heat transfer. The effect of this insulation must be calculated and quantified to ultimately establish an appropriate heat balance.

Heat energy is required to melt the casting alloy in the furnace. Some of it is absorbed in the die and die cooling system; some is lost to the ambient atmosphere. For the die surface temperature to be constant at a particular location, there must be a thermal energy balance in the die casting die.

Therefore, Heat input = Heat losses + Heat absorbed.

The production rate is limited by the heat load per cycle, which is determined by the specific and latent heat because they define the relationship between heat contained in the material and its temperature. Using the specific and latent heat values given earlier, the total heat released by the casting alloy between injection and ejection temperatures (heat load) can be calculated by the following equation:

$$Q_c = mc(T_i - T_e) = L_f$$

where

Q_c = Heat content per casting cycle
m = Mass of casting
c = Specific heat of casting alloy
T_i = Injection temperature
T_e = Ejection temperature
L_f = Latent heat of fusion

The rate of heat that is released by a particular net shape can be calculated by multiplying the production rate in shots per hour by the heat contained in each shot mass. This is the heat exchange that must occur by conduction through the die steels and cooling medium and the surroundings.

In practice, the production rate is set during cost estimating. Even though this rate is somewhat of a guess, for economical reasons, it determines the minimum target for quantitative thermal calculations. Of course, the maximum possible production rate should be calculated which also must include dry cycle machine time, injection and solidification time, die and core movement time, die spray time. Thus, the heat input rate can be calculated by the formula:

$$Q_r = P_{qc}$$

where Q_r is the heat load or the thermal work required from the die, and P is the production rate in cycles per hour.

Heat flux is the conversion of the heat load distributed over the entire surface area of the net shape being cast. This is a more meaningful calculation that is sometimes called the heat intensity. The heat flux caused by the superheated liquid metal applied at the die steel/liquid metal interface can be calculated by this equation:

$$q = Q_r/A_s$$

where q is the heat flux in btu/hr/sq.ft., and A_s is the surface area of the casting.

Then, the preceding equations can be combined to compute the heat flux as follows:

$$q = Pmc(T_i - T_e) = L_f/A_s$$

Heat flux is analogous to pressure since it provides an understanding for the heat loading from the superheated liquid metal on the surfaces of the die that come into contact with it.

Therefore, it is important to study the things that affect die surface temperature which are:

- Casting size and shape
- Production rate (cycles per hour)
- Casting ejection temperature
- Die construction (components and fits)
- Die materials
- Choice of cooling medium, temperature, and flow rate
- Die spray and air blow—flow rate and duration
- Size of thermal channels
- Location of thermal channels as related to specific casting details

A primary strategy in die casting is to maintain the temperature of the casting alloy above the liquidus during cavity fill. Predictable die surface temperature becomes an important support for quality casting production.

Casting details to watch for as sources of cold problems are thin walls and distance from the gate location. This suggests a plan to retain die heat or taking steps to raise the die surface temperature. Overflows with a low surface-area-to-volume ratio act as heat sinks and many times are helpful.

Electric cartridge heaters are available to the die caster to add external heat, but are not popular in practice. In the past these heaters have not proved to be robust enough and caused mechanical nuisance problems that accounts for their lack of use.

One might think that an obvious method to increase die temperature would be to merely run more shots per hour. However, cold conditions are not usually consistent for the whole casting. Therefore, heat problems might crop up where there were none at the slower production rate.

Heat-related problems occur when the casting alloy is too hot at the end of cavity fill, and even more often when the die temperature is too hot. Temperatures at ejection before die spray is initiated that are above 500–550°F are considered too high, except at the biscuit.

Details that are significantly more massive than the rest of the casting, such as heavy bosses or thick walls, especially

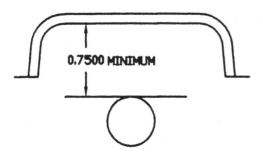

MINIMUM DISTANCE FOR COOLING CHANNEL

Figure 6

when a converging heat condition exists, are prime candidates for quantitative calculated thermal analysis. Experienced die casters can spot these structures and describe them as the last place to solidify. This is where to look for shrinkage porosity.

Certainly, the location of the cooling channel is key to dealing with such segments of the casting that run too hot. The excess heat must be extracted from the involved die component without adverse effects.

Of course, locating a cooling channel close to the hot spot is effective. However, caution must be exercised not to place it too close to the die surface because the thermal shock, especially if water is the cooling medium, is so destructive to the die steel that a crack can easily develop between the cooling channel and the die surface. If water seeps through the crack during cavity fill, steam will form that will eventually generate gas porosity.

The rule of thumb is to keep all cooling channels at least 3/4 in. away from any steel surface at the interface with the casting alloy. Figure 6 is intended to explain this spacing. It is not the center of the channel that is the focus, but the point on the circumference, usually referred to as the top, that is closest to the cavity surface. Even though hot oil removes heat more gently, it is well to also use the 3/4 in. rule with this medium as well.

As with the metal feed system, the hydraulic formula of $Q = AV$ applies to sizing the thermal channel. Reducing the diameter increases the velocity and vice versa. The flow rate of the cooling medium within the channel has a profound effect upon the transfer of heat from the casting to the die component, and finally away from the die.

Build up of scale in thermal channels is another factor that must be considered with cooling channel size. Most water contains minerals such as iron that combine to build up on the inside surface of the cooling channel. Waterline efficiency can be reduced as much as 40% with a scale of 0.005 in.! Therefore, while it is not an easy or popular procedure, water lines should be tested by running water (not air) through each circuit when the die is in the tool room for maintenance. The testing should also include measurement of maximum flow rates which should be in a range of 5–10 gallons per minute to be acceptable.

Usually, a minimum diameter for a through channel is considered to be 1/4 in., but 3/8 in. is better for water. Hot oil requires an additional 1/16 in. because of its different viscosity.

Another type of cooling channel, in wide use in die casting dies where it is not possible to machine a through line, is a fountain, also called a preculator or cascade, and is described here. The cooling medium is introduced through a hollow tube inside a channel that is machined into the die steel. Sometimes, though, a baffle strip, usually made of brass, is fastened

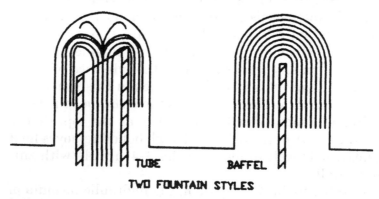

TUBE   BAFFEL

TWO FOUNTAIN STYLES

Figure 7

at the center of the channel and the medium flows from one side, over the top, to the other side. This design is frequently used where a series of this type of cooling is necessary. Figure 7 illustrates both types.

Since the direction of the flow changes in a tight bend, the minimum inside diameter of the tube is recommended at 5/16 in. even though much smaller tubes are commercially available. With smaller diameters, restriction to flow of the cooling medium can be expected and the cooling potential is less efficient. The equivalent hydraulic area between the outside diameter of the tube and the inside channel should be calculated to equal the area of a 5/16 in. diameter circle that represents the inlet area of the channel or pipe that carries the cooling medium to the fountain.

With the baffle style, the equivalent minimum hydraulic area is suggested for both sides. Figure 8 illustrates this principle. Equal inlet and outlet areas minimize any restriction to flow of the cooling medium except for the drastic 180° directional change.

As critical as temperature and heat are to the quality of die casting production, they are often given too low a priority because they are invisible, have no odor, and certainly cannot be felt at the operating level. Of course, thermometers and pyrometers may be used to measure temperatures at certain points in the casting cycle, but they are awkward to use. With convenience in mind, hand-held infrared guns that make regular temperature monitoring a reality are now available at a reasonable price. The latest upgrade here is continuous monitoring via a computer link that allows print outs in hard copy or downloading of electronic files for more comprehensive die surface temperature management. It is important, to adjust to the proper emisivity as recommended in the literature that comes with them.

Process variables are affected by many conditions which is the reason that the die casting process is considered by some managers to be so unpredictable and difficult to plan. This also justifies constant monitoring to control the process.

Get control of the metal temperature in the holding furnace or metal launder for a leg up on the production of

THIS CIRCLE REPRESENTS INLET AND
OUTLET AREAS TO FOUNTAIN

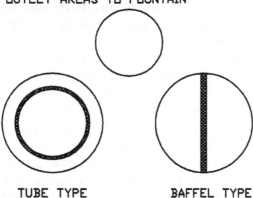

TUBE TYPE   BAFFEL TYPE

AREA ON EITHER SIDE OF THE
SEPARATION EQUALS THE EQUIVALENT
INLET AND OUTLET AREAS

TYPICAL SECTIONS THROUGH FOUNTAINS

Figure 8

high-quality castings. As with everything else, there are high
and low limits, but the range is broad enough to easily control.

The reason for the die caster's interest in metal temperature is that the objective is to keep the metal liquid during
cavity fill.

Die surface temperature is usually critical because the
casting alloy must travel over and come into contact with
the surface of the die steels as it moves from the gate to the
extremity of the cavity.

Since most die castings are shells with nonuniform wall
thickness, it is desirable to maintain the temperature of the
liquid casting alloy above the liquidus during cavity fill. In
this case, this is an easier task when the metal enters the gate
at a higher temperature.

Sometimes, castings with thick sections, heavy walls
or massive details are prone to shrinkage porosity. Here, it
is better to inject the metal at a lower temperature to
minimize the opportunity for volumetric shrinkage when

the temperature is at or slightly below the liquidus at the end of cavity fill. This usually involves reducing the injection plunger velocity.

Why is the plunger velocity of such interest? Most serious die casters today monitor both the slow and, especially, the fast shot velocity because the cavity fills during the fast shot phase. Thus, the temperature of the alloy at the end of cavity fill can be more easily controlled by manipulating the fast shot velocity.

The production rate indirectly affects so many of the process variables that it must be included in here although it has no direct bearing upon casting quality. For purposes of this discussion, the mechanical functions such as closing and opening the platens and extracting the shot will not be considered.

The production rate is slightly affected by the pouring time, but this portion of the production cycle impacts the injection temperature of the casting alloy.

A major element of the cycle is the dwell time which is manipulated so that the metal, which ideally is at the liquidus at the end of cavity fill, rapidly drops so that the casting is comfortably in the solid state by the time the dies open. This ejection temperature of the casting determines its initial structural integrity.

Of course, the surface-area-to-volume ratio of a specific casting determines the limits of the dwell time, but in many cases, the biscuit is the controlling factor because it has the highest mass in the shot.

External cooling via a water-based lubricant sprayed upon the hot die surface immediately after ejection in preparation for the next shot has a profound effect upon the quality of the casting produced. First of all, it must be understood that a major reason for spraying a water-based lubricant directly onto the cavity and core surfaces of the die prior to each casting cycle, is to keep the casting from sticking in the die after it has solidified.

However, another important effect is rapid cooling of the surface, whether intentional or not. When intentional, external die spray is utilized to cool only those surfaces whose shape is difficult or impossible to cool internally with a water channel, such as long thin cores with high heat concentration due to high volume-to-surface-area ratio. Despite the importance and almost excessive use of die sprays, their heat removal characteristics are not well documented nor understood.

For example, the mechanism of various parameters and deposition methods are merely part of the mystery. With this in mind, studies have been conducted to reveal that the temperature of the lubricant or coolant does not effect the cooling of the die surface and that most heat removal occurs during the first second of die spray. Application pressures are normally in the range of 60–80 psi, but do not significantly affect die neither cooling nor lubricant deposition.

Average operating surface temperatures vary from 350 to 600°F and, in this range, the dynamics of the interaction between spray droplets and the die surface is very complex. Remember this is an average temperature which fluctuates between 850° and below 350° as described by Fig. 1 in this chapter. For the spray to effectively cool the die surface and for the lubricant to be deposited, the droplets must contact the die surface. Then the lubricant carrier, usually water, must boil away.

The contact with the die surface is referred to as wetting. At high surface temperatures, a layer of steam is created and droplets cannot wet the die. At temperatures below 400°F, the die surface is relatively cool after spraying and little boiling occurs. The heat removal mechanism is primarily convection. Between 400 and 700°F, the surface is hot enough to boil water on contact and to remove heat through this phase transformation. For temperatures above 700°F and below 1100°F, the heat transfer decreases because of the layer of steam beginning to form at the surface. The steam layer is fully developed above 1100°F, which completely separates spray droplets from the surface.

The Leidenfrost phenomenon refers to the wetting temperature of the lubricant which is normally in the range of 300–350°F. This means that if the average die surface is above this temperature, the lubricant will sizzle on the die and not wet the surface, which seriously detracts from the lubricity. Some lubricants demonstrate a wetting ability in the 400–600°F range and these should be used for aluminum casting alloys that require higher operating die surface temperatures.

The Leidenfrost point is affected by variables such as droplet size, velocity, and roughness of the die surface. If the die has not been cooled to this point, no lubricant will be deposited and little heat can be removed from the die surface. Therefore, water is often sprayed on the die surface before the lubricant to cool the surface to this point so that the droplets can make contact.

The ratio of water to lubricant is important since this ratio determines the lubricity of the application. Normally this ratio is between 80:1 to 50:1, but sometimes is seen as low as 30:1 in especially difficult situations.

As a result of the general lack of understanding, external cooling is used too often to provide too much of the cooling (>50%) for the die. It becomes an integral part of the thermal system for too many die casting dies. It is recommended that this role of die spray be minimized to merely a spurt, which can only be accomplished with carefully analyzed and designed internal cooling systems.

In its role as part of the cooling strategy, the flow rate of the release agent through each spray nozzle is not as important as the duration of spray. A flow rate of less than 1 quart per minute is satisfactory and duration time of application should be less that 5 sec unless proper thermal calculations are made. The usual method is to adjust the die spray by trial and error until the die operates satisfactorily, but such practices are what have given die casting its reputation for being unpredictable.

Internal thermal design should be versatile enough to utilize two cooling mediums. The most widely used is water which is recirculated at temperatures in the 85–100°F range; the other is calorific oil which is used in the 200–300°F range. However, the temperature of the cooling medium is not as critical as the flow rate but must be controlled in tandem with it.

The flow rate of the cooling medium has a significant effect upon the die surface temperature, which directly determines casting quality. Therefore, this is the facet the experienced professional will use to deal with quality issues because it can be calculated and specified into the operating window. It is more important than the temperature and often is effective in transferring heat at rates from 1 to 5 gallons per minute through each channel.

This writer has observed that the capacity of some central water recirculating systems are under-designed so that when the run time of the total casting department exceeds 65% and approaches a more desirable level of greater than 90%, the recirculating pumps cannot keep up and even 5 gallons per minute becomes a challenge. It is very difficult to quantitatively design a proper recirculating system without having first calculated the heat transfer requirements of all of the dies. Thus, the effective system will usually be one that is over-designed by guess and by golly.

Hot oil systems are designed to operate on a single die or zone and are more suitably sized because they are isolated from the indefinite aspect of the other casting machines.

Every one of these conditions can directly impact casting quality by itself or in tandem with one or more of the others. All are mechanically controlled by valves, pistons, pumps, switches, etc., which can deviate from the desired setting or be easily changed. Therefore, as long as humans have access to the controls, regular deviations can be expected. It is the combination of thermal process variables described here that must be addressed by a well-designed operating window. A broad window will successfully tolerate a wide range of deviations and the die will be easy, even for beginners, to operate, and vice versa.

Remember, die casting is a thermal process and heat energy is the engine that drives it, so for the best results extreme care must be taken to understand this important principle.

10

Designing the Value Stream

Since high-pressure die casting is an extremely capital-intensive process, it behooves the astute manager to obtain the highest yield possible from each expensive die casting machine. The objective of course, is to maximize the return on investment.

Two things that detract from this goal are downtime caused by tooling or equipment problems, and defective or scrap castings. It has been said that a 5% reduction in scrap will double the profit in a die casting operation. This calls for a problem solving procedure to minimize casting defects. If left to a cut-and-try or seat-of-the-pants technique, the die casting process is extremely unpredictable and difficult to manage.

This brings to mind some quality events that occurred early in this author's career that point out inconsistencies that occur without a strategic system. On difficult jobs, it was not unusual to pass 5000 salable parts on one shift, only to see 1000 on the next. How can there be any consistency to quality or through put, under such circumstances?

Several modern methods are helpful in defining the cause and ultimately curing the problem to eliminate the defect or defects. They are all based upon historical and subsequent data analysis, focus upon the process rather than inspection of product, and normally call for a severe adjustment of corporate culture (Crossen, H. M.). Some of these strategies are reviewed here because they each quantitatively define the defect and determine the cause. Effective mitigation is the final result that is usually dramatic.

Lean technology can best be defined by five principles. (Nicol, 2003):

1. Value—is what the end user of die casting really wants. It is what sets the end product apart from competition and establishes its true quality. It is imperative that every player in the supply chain understands this requirement.

2. The value stream—must be designed to eliminate waste on corporate resources. In this context, casting defects represent scrap or unsalable product that is the waste on which to concentrate. Lean technology is relentless in pursuit of perfection. Its resoures are totally integrated into the value system.

3. Flow—must be continuous and the profitability that results from increasing inventory turns, reducing defects, and other disciplines that are unrelated to this subject is a great motivator.

4. Pull—is a function of quick turnaround time and reduced inventory uncontaminated by defects. The customer literally expects to pull finished, defect-free castings from the die casting cell. It is like having the boss press a button to start all of the manufacturing operations until they produce the quantity of castings required by the pull.

5. Perfection—is realized when the end user recognizes that the value stream works so well that on-time defect-free parts can be pulled.

Lean technology is simplified here for brevity and is not easy to accomplish. However, the discipline that it inspires can be a very effective mitigation of casting defects.

Six Sigma is a disciplined data driven method for eliminating defects that is directed toward six standard deviations between the mean and nearest specified limit (George Group, Internet Nov. 2003). It statistically describes how the die casting process is performing. The objective for Six Sigma status for a specific die casting process is to produce no more than 3.4 defects per million opportunities. A Six Sigma defect is defined as anything outside of customer specifications. A Six Sigma opportunity is then the total number of chances for a defect.

An improvement system for an existing, nonconforming process (or die casting die in our case) that defines, measures, analyses, improves, and controls, is an important facet of Six Sigma culture. A strategy for development of a new process that will perform at Six Sigma quality levels calls for designing, measuring, analyzing, and verifying.

Unfortunately, there is a gap between the Six Sigma acceptable status of 3.4 defects per million and this writer's experience with the high-pressure die casting process, which is approximately 5,000–10,000. In order to reduce the defect rate, it is necessary to tighten the control function, which requires closed loop adjustments of plunger velocity and pressure or data-based monitoring that prevents castings produced outside specified limits from entering the value stream. These systems are explained in the chapter on process control.

Design of experiments (DOE) is a more common method for minimizing defects; determines the root cause that establishes accepted limits for appropriate variables. As with the other disciplines, a consensus team composed of critical players must brainstorm the root cause of defects. The chosen variables are then inserted into a series of experiments in which actual parts are die cast and subjected to customer specified quality standards. Thus, the DOE identifies the root cause of the defect and sets acceptable limits by experimentation rather than by analysis. This does not necessarily mean that optimum operating limits are achieved, only that customer specifications have been met. The best operating window is set only for parameters that can be adjusted within the casting work cell.

If improvement in runner and gate, venting, or internal thermal system (cooling channels) is suggested by the consensus team, it needs to be analyzed and calculated physical revisions made to the die. When this step is added to DOE, it is possible to set up the die casting process to *exceed* customer specifications and the Six Sigma gap starts to close.

The practical approach is to clearly identify the specific problem that is causing the defect, and then proceed to solve it with a quantitative strategy. This chapter attempts to suggest some practical methods.

Control charts are used by die casting firms that expect to survive in an extremely competitive environment to control, monitor, and *improve* the casting process. Root causes for defects are always present because they are attributed to shot end repeatability, casting alloy, time, or temperature. If physical details such as sleeve or gooseneck, runner, gates, vents, and thermal channels are properly designed, only minor adjustments to the process are necessary to restore it to specified control. If a trend is noted that suggests a deviation, it should *always* require only a minor change. Too often, technicians will over correct by making too great a revision, which then causes the process to dramatically shift in the other direction.

A control chart is a graph with limits called control lines that define the upper control limit (UCL), the central nominal line, and the lower control limit (LCL). Its purpose is to detect any deviations in the process that depict abnormal points of collected data. Without closed loop process control or automatic side tracking, the points should be plotted in real time so that the die casting technician can make an immediate adjustment. The adjustment should be recorded, and the cause of the drift and what action returned the process to a state of control noted.

For broad control, these points are usually averaged into subgroups to minimize the quantity that must be plotted. It is when some of these points fall outside the upper or lower control limits that the root cause needs to be investigated and appropriate action taken to keep it from recurring. The quality term for this preventive action is continuous improvement. Defects can only be prevented if correction is made prior to the deviation exceeding the limits. If not, scrap is generated.

Two types of data are used for control charts. Data such as time, temperature, or velocity that is based upon measurement are called indiscrete or continuous data. Other data based on counting, such as quantity of castings produced or number of defects, are known as discrete values or enumerated data.

The X bar and R control chart is commonly used to control the high-pressure die casting process. The mean value (X) describes changes in the process, while the R portion shows any deviations or process dispersions from the mean. The analysis of process data requires calculation of the control limits. Any X data will have both a mean (mu or X bar), and a standard deviation. Most of the data (99.7%) will fall within + or −3 (Sigma) of the mean. Mathematical calculations that utilize specific formulae are necessary to establish UCL and LCL limits. These calculations will not be explained here since expert texts are available elsewhere.

Since both X and R are illustrated at the same time, the control chart is a very effective method for checking abnormalities within the process. If they are charted during actual production, a casting problem can be announced in real-time mode.

For the process to be in control, it is important that X bar deviations must be random and, of course, lie within the UCL and the LCL.

Some suspicious data that would not be considered as random are:

- Too many points in middle third of range
- Too few points in middle third
- Runs of seven data points above or below X mean line
- Cyclic patterns that describe a trend up or down
- Violation of control limits

Process variables that could be plotted are:

- Slow shot plunger velocity
- Fast shot plunger velocity
- Biscuit thickness (cold chamber)
- Die surface temperature

- Holding temperature of casting alloy
- Cavity fill time
- Accumulator pressure
- Rise time
- Cycle time
- Intensification pressure
- Dwell time
- Die spray time
- Flow rate of cooling medium

Most dies, however, will require the gathering and plotting of no more than five variables, depending upon the type of defect and the shape of the casting produced.

A Pareto chart is sometimes useful to determine the frequency of defects in order to concentrate improvement efforts where the potential for improvement is the greatest. This is a classical method, which determines the vital few defects that stand in the way of lowering rejected parts-per-million performance.

Internal defects are sometimes caused by foreign inclusions in the casting alloy. There are several types from different sources and are outlined here to alert the die caster of their possible presence along with some suggestions to deal with them.

Oxides are formed when oxygen is introduced. This can occur in melting or alloying. The source can be from floor sweepings or products of combustion, which is exacerbated by humidity and turbulence. Back scrap consists of scrap castings and runners that have a thin skin of the base metal and oxygen.

The rate of oxidization doubles for every 20°F rise in the temperature of the liquid metal bath as it is superheated for die casting. The oxides formed morph into particles or skins.

Aluminum oxide cannot be re-melted back into aluminum because the soft gamma form converts to the hard alpha form at elevated temperatures above 1400°F. Corundum is the hard alpha form. An example of corundum is illustrated in Fig. 1 by a saw cut through an aluminum casting (Walkington, 1997).

Refractory materials from the furnace lining can be degraded into inclusions in the form of corundum. This

Figure 1   Corundum.

eventually forms a hard spot, a casting defect that can adversely affect machinability.

Carbides come in particle form from scrap that contains oils; they also contribute to hard spots.

Halide salts are sometimes introduced in particle form into the bath via reaction of fluxing products.

Inclusions in zinc alloys are composed of aluminum and iron. A hard spot is formed when the aluminum in the zinc alloy dissolves the iron from the gooseneck or holding pot, a good reason to avoid a cast iron pot. Temperatures above 820°F increase the incidence of inclusion.

Scratch marks during buffing and excessive cutting tool wear in machining are the result of this defect.

Occurrence of inclusions may be minimized by allowing sufficient settling time for the liquid metal bath after cleaning of the furnace. Frequency and procedures are covered in the chapter on metal handling. Excessive melting temperatures should be avoided. Good maintenance is essential to prevent air leakage into the furnace and to hold down oxidation. Back scrap should be charged in a ratio of no more than a 3:1 mix with virgin alloy.

Degassing to remove hydrogen picked up from the atmosphere, especially in humid weather, is recommended. Inert gases of argon or more economical nitrogen are suggested. If a rotor is used, smaller sized hydrogen bubbles with a greater surface area will be purged.

Most defects are caused by deviations in process variables. Therefore, in order to quantitatively identify the problem, the operating window must be examined. (Photographs of defects are not included here because they are easily accessible in the NADCA texts.)

Cold shut—sometimes called poor fill or cold lap, or even lamination—is the most visible defect, as it can be readily seen on the surface of the casting. As its name implies, it is caused by temperatures that are too low. This defect suggests that temperatures during cavity fill are not being managed properly. This common defect in castings produced from all casting alloys is depicted in Fig. 2.

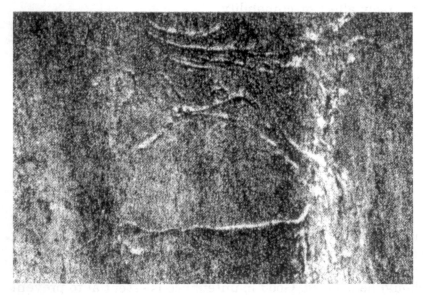

Figure 2 Cold shut defect.

The temperature of the metal in the holding furnace or crucible is the most likely root cause, outside of the cavity filling process. Chapter 4 provides some guidelines that define the recommended range of allowable operating temperatures for the different casting alloys.

The temperature of the die, especially in the area of the cold shut, must be checked directly after the shot is ejected and prior to the application of die spray for the next cycle. Generally, temperatures below 350°F for zinc, and 450°F for aluminum and magnesium alloys are too low.

Excessive die spray can create surface impediments to the flow and can cool the die surface too much to generate cold shut. Usually, however, a long cavity fill time resulting from a fast shot plunger velocity that is too slow is the more likely cause. This allows too high a portion of the metal to solidify before the cavity is completely filled.

Some suggested allowable limits for percentage of solidification during cavity fill are suggested in Table 1 for different casting alloys with usual wall thicknesses (0.03–0.04 in. for zinc, 0.08–0.125 in. for aluminum, and 0.05–0.08 in. for magnesium).

Table 1 Percentage of Solidification During Cavity Fill

Casting type	Zinc	Aluminum	Magnesium
Highly decorative	0–5%	0–10%	0–7%
Decorative	5–10	5–15	5–10
Functional	10–15	15–20	10–15
Functional (min. porosity)	—	5–15	5–10

Many times excessive cavity fill time is the result of insufficient gate area. In this case, when an attempt is made to fill the cavity faster by increasing fast shot plunger velocity, the gate speed becomes excessive and causes other quality problems. Flow in superheated liquid metal streams is too turbulent and creates uncontrollable swirls that exacerbate a cold shut defect.

Dimensional distortion can be caused by product design features such as heavy sections, thick ribs attached to thin walls, or just thin walls with no ribs. Thermal gradients between different regions of the shape to be cast are always involved in this defect. However, there are also some process variables that can be the root cause of distortion as well.

Too great a range between die surface temperatures within the same die cavity cannot be tolerated if any extraordinary degree of dimensional tolerance is desired. To minimize this defect, analytical management of the temperatures of the casting alloy as it travels through the metal feed system is essential.

Die surface temperatures that are excessive will retard solidification after the cavity is filled so that the casting is too ductile for ejection. This plastic condition does not contribute to dimensional stability.

Since it is advantageous to reduce the temperature of the casting at ejection, another way to deal with this problem is to increase the time that the dies remain clamped together (dwell time) to give the metal more time to solidify. This tactic, of course, also reduces the production rate.

Heat depressions, sometimes called sinks, on the surface of a casting are usually the result of unbalanced die surface temperatures where the die temperature is hotter on the side of the depressions than on the other side of the casting wall. The depressed areas are located at the last places to solidify. The surface is smooth and usually has a frosty appearance. Many times, this defect is located on the opposite side from a massive or heavy feature of the casting. Since this is a heat problem, it will be made worse by excessive die spray that is applied to remove the heat from the die component. An example is shown in Fig. 3.

Many times this pattern can be found in a sunken detail of the casting shape that is formed by a raised die component because it is more difficult to remove excess heat from a standing feature in the die cavity. Sometimes the problem can be minimized by raising the temperature of the die on the other side of the wall by reducing the flow rate of the cooling medium. It may even be necessary to lower the die surface temperatures by slowing down the production rate.

It is also possible, however, that the fill strategy is inefficient, and in this case, an increase in flow rate will solve the problem. This can be accomplished by either increasing the

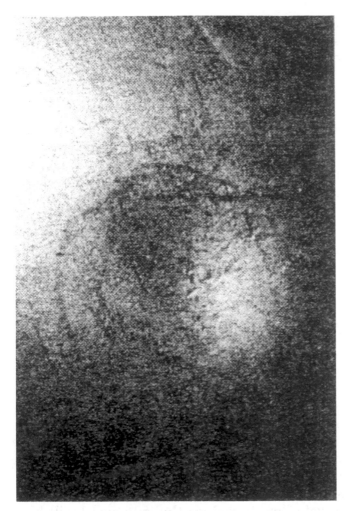

Figure 3 Heat depression or sink.

diameter of the plunger tip or the fast shot velocity, which will decrease the cavity fill time to maintain the superheated temperature of the casting alloy.

Another solution for eliminating this defect is to use tungsten-based steels that have a greater heat transfer coefficient than H-13 die material. A reduction in metal injection temperature obviously will help as will a reduction in the nozzle temperature if the hot chamber casting method is used.

This condition can also be helped by increasing the accumulator pressure as well as the intensification pressure. The appearance of heat sinks can be diminished by slightly bowing a flat surface, but this requires the customer to agree to an engineering change... a most difficult task outside of simultaneous design of product and die casting die.

Solder may not be a defect unless the rough finish is not acceptable, but it offers an impediment to producing quality castings. It is described as casting alloy adhering to the casting die. It is heat related, therefore the cause could be metal injection temperatures that are too high as well as die surface temperatures that are too hot. Thus, improved cooling of such die surfaces can be accomplished by adjusting flow rate of the cooling medium through adjacent cooling channels or possibly changing the core material to a tungsten-based steel alloy that displays greater thermal conductivity than H-13. The frequently encountered condition is depicted in Fig. 4.

Soldering usually occurs when the metal stream impinges, at high velocity, upon an obstruction in the die such as a steel core. The protective coating on the core is washed away so that the casting alloy may then bond to the die steel. When the casting skin stays with the core, the metal that sticks (solders) to the die member is light in color and might even shine. Since aluminum casting alloys dissolve the iron in the die steel, 1% iron is alloyed into the aluminum to slow down this activity. Therefore, if the alloy is low in iron content, soldering occurs more quickly (Chu et al., 1993).

Since the condition is exaggerated by high gate speeds, increasing the gate area to reduce the gate speed will help to minimize the defect.

In addition to dissolution of the substrate elements of the die material and pitting corrosion by superheated liquid aluminum, other patterns are apparent in the die substrate that are caused by diffusion of aluminum and silicon elements. An intermetallic compound (Al–Fe–Si, Al–Cu–Si) is formed.

Diffusion is the spontaneous movement of atoms to new sites in a material. Processes depend upon many factors, but mainly on gradients of concentration of related atoms and temperature. The diffusion rate increases with

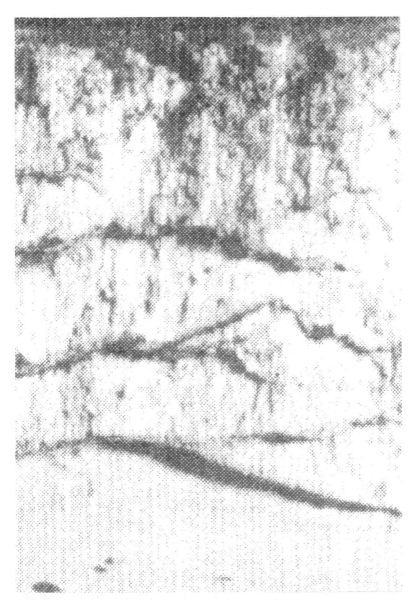

Figure 4 Solder.

temperature since higher temperatures increase atomic motion. A laboratory section is examined in Fig. 5 with a distribution profile of the soldering layer.

Figure 5 Distribution profile of soldering layer.

Eventually, the virgin steel is exposed to liquid aluminum and suffers from thermal corrosion and diffusion. Figure 6 brings perspective to distribution of the elements with the layer of solder.

Porosity is inherent in the high-pressure die casting process. It is one of the primary defects that can be found in all die castings and is in distinct contrast to the fine dense grain structure that makes die casting a desirable commodity. This defect is more common when castings are produced in aluminum alloys. The turbulent nature of the process makes it difficult to control porosity, so it is a major concern to both die casters and users of die castings.

These voids are usually quantified by size and quantity to establish acceptable quality standards. Porosity is found in different degrees of intensity that are a function of the casting shape and the methods used to produce the part.

This is of such concern to both producers and users of die castings, that several methods (CSIRO, 1992; NSM/OSU, 1991) have been scientifically suggested to determine the extent of porosity in a particular casting. They are listed here

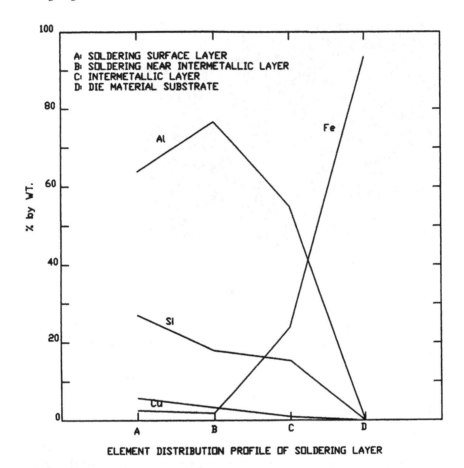

ELEMENT DISTRIBUTION PROFILE OF SOLDERING LAYER

Figure 6

for information only; it should be understood that each has limitations:

- Archimedes' method for density measurement
- Radiography
- Ultrasonic attenuation
- Metalographic examination
- Vacuum fusion to determine contained gas contents

While the methods listed above are very precise, more mundane but practical methods are used because of cost and time involved. Some of these are sectioning to expose

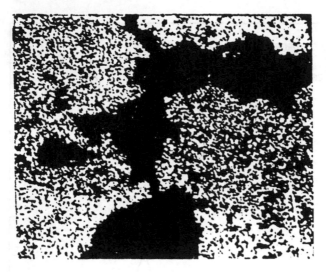

Figure 7   Shrinkage porosity.

the porosity, pressure testing, radiography (also listed above), and subjecting the casting to the functional requirement that it is designed for.

There are two basic types of porosity and each is the result of very different circumstances. These types are shrinkage porosity and gas porosity that are described below.

Shrinkage porosity occurs because, as liquid metals solidify, they reduce in volume. Therefore, under the right conditions, voids occur where a usually massive detail (high volume/low surface area ratio) in the cast shape is literally torn apart to accomplish this volumetric shrinkage. In such a detail, the grain structure is more likely to be dendritic or tree-like. The tearing occurs because the supply of metal to that detail has been exhausted. These voids are called shrinkage porosity and can be found at the last place in the casting to solidify. In the trade, this is called the "last place to fill". The defect has ragged surfaces, which are the result of tearing and the dendritic structure.

Shrinkage porosity has a distinctive appearance characterized by jagged edges and sharp corners as a result of the tearing action. An obvious strategy to reduce the mass can be initiated by the product designer who recognizes such a situation early on in the design procedure. A laboratory section is offered in Fig. 7.

The solution related to the die casting process is to provide sufficient gate area to continue to feed the area after the rest of the cavity is completely filled. This is done during the intensification phase of the casting cycle when the pressure upon the metal is normally tripled.

Another method to deal with shrinkage porosity in aluminum alloy castings, is to consider using the higher silicon content alloy specified as 413, which has a freezing range of only 20°F between liquidus and solidus rather than the 100° range for A380 alloy. This alloy provides less opportunity for volumetric shrinkage, which must take place prior to solidification.

It is always a good practice to check the chemical composition of the casting alloy since shrinkage defects occur less often if the alloy has more resistance to hot cracking. This feature diminishes if the iron content is allowed to fall below 0.8%, the zinc content above 4%, or the magnesium content below 0.3%.

In the event that all other efforts fail to cure the problem, a more heroic strategy is also available. A densifier pin can be installed in the die that is designed to apply external pressure directly at the point of the shrinkage void. This is done by compacting the still slushy metal into the void so a dummy appendage must be added to the casting to supply the extra metal.

A leaker is a derivative of shrinkage porosity that can be found in very massive areas of a cast shape in the form of a long path through which gas or liquid can escape so that the casting leaks. A leaker path is depicted in Fig. 7, a microscopic 500× photograph where the as cast surface is to the right and the shrinkage porosity is at the left (The Ohio University).

Centerline porosity is another form of shrinkage porosity that forms at a neutral thermal axis within the wall of a casting during cavity fill. Many times it is microporosity and almost invisible to the eye, but in some cases it can be a cause for knashing of teeth.

The hotter the die surface temperature, the nearer the thermal axis will be to the surface. Thus, it behooves the die caster to balance die temperatures in mating die halves as closely as physically possible. This will locate the thermal axis nearer to the center of the wall thickness. This condition is depicted in Fig. 8.

Gas porosity can be identified by the relatively smooth surface that it presents during examination. The void is actually an air bubble that becomes encapsulated by the liquid casting alloy during cavity fill. The entrapped air can originate at any of several points in the metal feed system. An example of what typical gas porosity looks like is presented in Fig. 9 as a sectioned casting.

If the surface of the pore is clean and, possibly, shiny, the cause is probably over spraying and water left on the die surface prior to injection, or from entrapped air in the runner.

In addition to reducing the resistance to tensile stress or impact, these air voids also are a major cause for casting rejection when they are opened up in machining. In such cases, they can damage cutting tools or cause functional

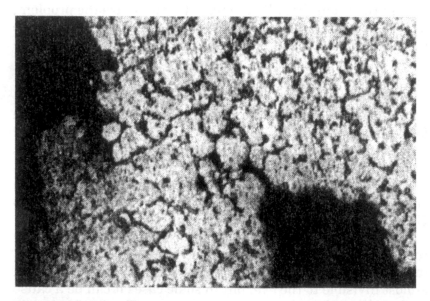

Figure 8 Centerline porosity.

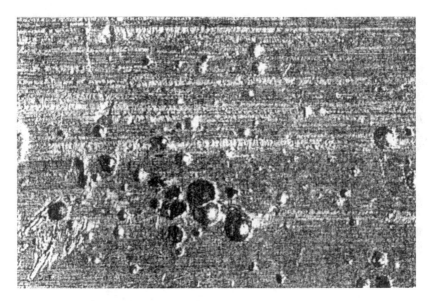

Figure 9 Gas porosity.

difficulties on the machined surfaces. Many times porosity cannot be identified as simplistically as this discourse might suggest, so a photograph of a casting section that includes both shrinkage and gas porosity is included here in Fig. 10. The smoother group to the left is gas porosity and the more irregular ragged grouping at the right is, of course, shrinkage porosity.

The usual sources for entrapped air are:

- Gas formed in the metal as a result of poor melting and handling procedures
- Steam formed by heat from contact of the superheated casting alloy and moisture on the shot sleeve or die
- Hydrocarbons from burning of excess oil-based release agents
- Turbulence in the shot sleeve during the slow shot phase
- Splashing at the intersection between the runner and sleeve
- Poor runner design
- Poor design of cavity fill pattern

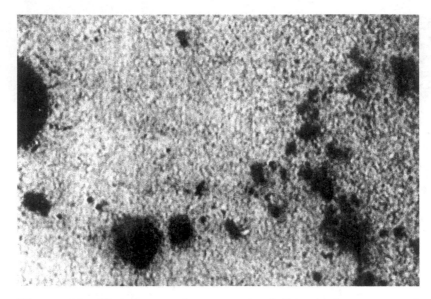

Figure 10   Shrinkage and gas porosity defects in the same casting section.

Gas porosity can be reduced in size and quantity, and it can be moved away from the critical area of the casting to improve the internal integrity of the casting.

Edge porosity is another variation that can occur at either the gate inlet or at the outlet to overflows or air vents. It is characterized by small pores (holes) that can be seen after the gate or overflow has been broken away. Sometimes, the skiving that takes place during the trim operation can smear over small holes to diminish their appearance. An example is shown in Fig. 11.

It is difficult to identify the cause of this defect because so many different elements are involved. If the condition is located at the gate inlet, between the runner and the casting edge, the holes can be the result of air entrapment that takes place in the metal feed system. It includes the shot sleeve and the runner. The slow shot velocity must be examined because excessive splashing during this phase can cause edge porosity.

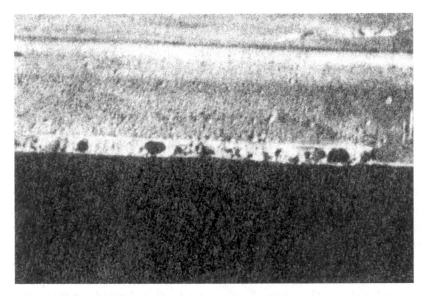

Figure 11 Edge porosity.

The design of the runner could be the cause if it does not constantly decrease in cross-sectional area as it approaches the gate location.

If the porosity is between the casting and an overflow, it probably is caused by a low die surface temperature and/or low metal temperature in a location that is remote from the gate. Premature solidification during cavity fill can be the cause. Also, slow gate speed as the metal exits the runner is an even more probable cause.

Many strategies are available to deal with porosity caused by air entrapment because of the multitude of conditions involved.

First, several degassing devices and methods are available to remove excess gas in the casting alloy. Melting and holding temperatures of aluminum casting alloys should also be controlled to minimize the potential to pick up hydrogen gas from the hydrocarbons from combustion of fuels.

Die spray and plunger tip lubrication should be used as sparingly as possible by the proper design of the thermal system in the die.

Turbulence during the slow shot phase while the sleeve and runner are gradually filled with liquid metal can be controlled by following the critical slow shot recommendations.

Turbulence in the runner system can be controlled by streamlining all sharp changes in direction and constant reduction in cross-sectional area as the metal travels toward the gate. Thus, the velocity of the metal stream will constantly increase so that the fastest speed is reached when the metal exits the runner through the gate.

The fill pattern of the metal streams as they fill the cavity should be carefully planned and the last place to receive metal must be accurately identified. It is important to exhaust as much air from the metal feed system as possible in front of the metal stream.

By locating the vent away from the place in the cavity to receive liquid metal, and sizing it so that the air will escape at approximately the speed of sound, the liquid metal can displace the air that was in the system before the pour is made.

Heat-related defects, caused by poor management of the temperatures of both the casting alloy and the die surface, are many times manifested as blisters. A blister is gas porosity that is located near the surface of the casting. Entrapped gas will gravitate to the highest die surface temperature. During solidification, the gas is under very high pressure—up to 15,000 psi. Immediately after the casting is ejected from the supporting die steels, the temperature of the casting alloy is high enough (500°F) to keep the casting in the plastic range of deformation. It is during this phase that the soft metal easily expands to form a blister. Such a defect is illustrated in Fig. 12.

Cavitation, is mentioned in Chapter 6 on metal feed systems, is also considered a defect because it causes a type of erosion to the die steels that forms bumps in the casting that are the negative of the pits in the die steels. Of course, this defect affects 100% of the production so it is a problem that has to be solved immediately. It occurs usually when casting metals of higher specific gravity such as zinc and lead. It is rarely seen in casting aluminum alloys, and never with

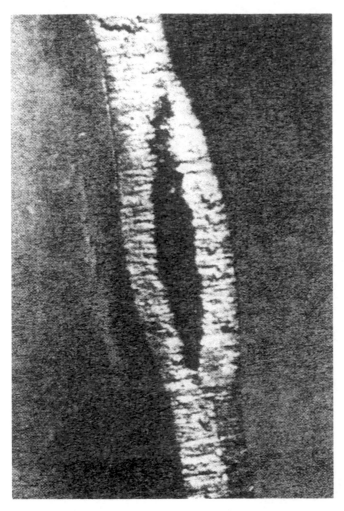

Figure 12 Blister.

magnesium. To illustrate this phenomenon, an early scientist, Brunton, took a series of high-speed motion pictures of the collapse of a bubble in water, which are presented in Fig. 13.

Note the asymmetrical collapse near the surface that simulates the high-velocity impact of jets and droplets in the flow of zinc alloy as it moves through the cavity fill phase. The cause is the resistance of heavier metals to directional

Figure 13 Collapse of bubble in water.

change. Cavitation normally happens in the runner, just upstream from the gate, and manifests itself slightly down-stream from the gate. The fix is to create a more gentle transition from one direction to another.

The trouble shooting chart in Fig. 14 may provide a help-ful reference to the causes and possible cures of many die casting defects. The defects are posted across the top, and

POSSIBLE CAUSE	COLD SHUT	DISTORTION	HEAT SINKS	SOLDER	EDGE POROSITY	SHRINK POROSITY	GAS POROSITY	DIE BLOW [SPIT]	BLISTERS	HOT CRACKING	DRAG MARKS	HARD SPOTS	EXCESSIVE FLASH	BENT CORES	CRATERS AND PITS	SINKS AND SHADOWS	INTERGRANULAR COROSION	UNSTABLE CREEP [ZN]	LOW DUCTILITY [ZN]
METAL TEMPERATURE	x		x	x	x	x				x	x		x						
DIE TEMPERATURE	x		x	x	x	x				x		x		x	x	x			
DIE SPRAY [LUBRICANT]	x		x	x						x		x		x	x				
CAVITY FILL TIME	x							x											
GATE SPEED	x		x	x				x		x									
WALL THICKNESS	x	x					x									x			
DWELL TIME		x																	
PRESSURE		x						x		x	x	x		x					
CASTING ALLOY			x				x	x		x	x	x					x	x	x
CAVITY FILL PATTERN			x					x				x							
PLUNGER VELOCITY																			
VENTING	x				x		x												
PLUNGER DIAMETER					x			x											
RUNNER DESIGN							x												
LOCKING FORCE									x					x					

Figure 14

the potential strategies to eliminate the defect are at the right. It must be noted though that this information is offered only as initial suggestions since actual die casting defects are not usually so simplistic.

11

Die Materials

The materials used for die casting dies are mild steel alloys, cold or hot rolled, that are either air, oil, or water hardened. These materials contain 0.3–0.4% carbon as well as chromium, molybdenum, and vanadium as major alloying elements. A most challenging application is in the production of aluminum die castings since over 70% of all die castings are from this casting alloy. The reference here is to the inserts that are the heart of each die. The bill of materials for each die design should specify the type of steel and its hardness for each die component.

Cavity inserts "see" the superheated liquid casting alloy and must be strong, tough, wear resistant, and able to withstand thermal fatigue. The metallurgical term usually applied to these characteristics is fracture toughness, which is a factor of the ductility of the steel. Thus, the material needs to be designed to minimize deterioration during repeated casting cycles that require it to dramatically expand and contract. The thermal fatigue mechanism is the same as that experienced when a wire is bent back and forth until it finally breaks. In die casting die steels, this behavior is called heat checking.

Heat checking is caused by the dimensional expansion that occurs as liquid metal at superheated temperatures and the rapid contraction when the casting alloy solidifies. However, throughout the casting cycle, the core or interior temperatures remain relatively stable, which concentrates the stresses at a thin layer at the interface between the casting alloy and the steel surface. This layer is normally only 0.06 in. thick.

The external die spray duration for lubrication and cooling has the greatest impact upon die checking because of tensile stresses at the thin surface layer. The relationship between these stresses and temperature is described in Fig. 1.

Therefore, the chemical content, grain structure and alignment, internal integrity, cleanliness, and heat treatability are key factors that must withstand the thermal extremes of the production cycle. It is more severe and prevalent with aluminum casting alloys that are injected at higher temperatures. It occurs with zinc, but it is rare.

The useful life of a die is governed by the fracture toughness of the material at the interface with the casting alloy. Normal die life for zinc dies is at least 1,000,000 cycles, but only 150,000 for aluminum.

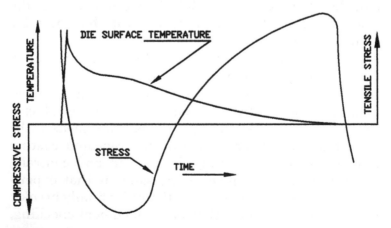

Figure 1

Theoretical die life can be enhanced by as much as 50% by compressing the gradient between the highest and the lowest temperatures that the die material experiences during each casting cycle. The highest temperature occurs when the casting alloy is aluminum and is typically approximately 900°F during cavity fill. Although this is not really controllable, the lowest temperature after external die spray of about 400°F can very well be controlled. If the lowest die surface temperature is increased to 600°F, the thermal stress will be dramatically reduced by 3/5 or 60%!

How can this be done? As previously stated hundreds of different die casting plants, every one operated their dies too slowly, save one, and that one broke all of the accepted thermal rules. Therefore, *run the die faster* so that the lowest die surface temperature will increase. Of course one cannot merely reduce cycle time and more shots per hour without the necessary preparations.

Dies run too cold except for one or two hot spots and the biscuit. Usually, it is quite possible to design sufficient internal cooling to balance the die surface temperatures within a reasonable range so that more shots per hour are produced. The die material will perform much longer, and just think what 10–20% greater productivity will do to casting cost!

Though heat checking is the most dominant failure mechanism in die casting, another is gross cracking. It is less common but more significant because it is unpredictable and usually catastrophic. The key to optimizing die life is to establish properties that discourage gross cracking and at the same time delay initiation and propagation of heat checking as long as possible.

Most cavity inserts are made of P-20, H-11, H-13. P-20 steel is used for die casting zinc, and H-13 is used for casting aluminum and magnesium in addition to zinc. In Europe, H-11 steel is extensively used for casting the same alloys. For brass castings, H-21 steel is used. All of the available die materials have limited die life because of creep rupture properties. H-11 and H-13 are regarded as multipurpose materials due to properties at high operating temperatures.

These die materials are usually electroslag remelted, but the better grades are remelted in a vacuum to control nonmetallic inclusions (cleanliness). The steel is then forged to reduce its volume by a factor of 5 to refine grain structure and to close any remaining voids.

P-20 steel is a low carbon alloy with 1.5–2.0% chromium, and small amounts of silicon, manganese, and molybdenum. This material has good toughness and excellent machinability. Unfortunately, this material is not adequate for the higher thermal stresses imposed by the higher temperature casing alloys. Therefore, it is restricted for use with zinc alloys only.

H-13 steel is a chromium type hot work tool steel used for die casting aluminum and magnesium, and also for some high volume zinc die casting. This material contains 5.0–5.5% chromium, over 1% molybdenum, 0.4 carbon, and almost 1% vanadium.

Maraging steel is a very low carbon, silicon, and manganese die material with relatively high nickel, cobalt, and molybdenum. Though not especially popular, it is mentioned here because it has high strength while still retaining good ductility. It is used frequently for welding H-13 die components since it is relatively soft, but the interface with the H-13 is very hard. This hard zone is tempered to a more desirable hardness at 1000°F for 3 hr, which also hardens the maraging.

Upon hardening at 900°F, maraging steel shrinks 0.0005–0.001 in. per inch in all three dimensions. This makes dimension calculations too complicated for popular use.

Machining of die steels to create the detail of the shape to be die cast is accomplished by the usual cutting methods of milling, drilling, etc. It is almost universally done under computer numerical control (CNC) by following pre-established tool paths that are imported from a 3D CAD model of the shape as designed by the product designer. Most cavity shapes of any size are rough machined (hogged out) prior to hardening for ease of removal of the mass of steel. This is done to avoid dimensional distortion that always occurs during the violent temperature gradients that the die material is exposed to during heat treatment.

Final machining is then done with specially hardened cutting tools applied to the steel in the hardened condition at a hardness of 44–46 on the Rockwell "C" scale. This procedure does not affect the die surface adversely, but is slow and expensive.

Electrical discharge machining (EDM) is considerably more economical since a mirror image of the shape is machined in a soft graphite electrode that is then transferred to the die steel. An electrical field is created in dielectric fluid between the electrode and the work.

The problem is that the process leaves an undesirable surface condition on the finished steel (Dorsch, 1991). The EDM'd surface consists of a resolidified "white" layer of as-cast and as-quenched martinsite plus a soft overtempered layer approximately 0.003 in. deep, just below the white layer. EDM literally melts away the material that would be machined away and then rapidly solidifies the thin surface of the steel after the desired shape has been thus formed.

Both layers degrade resistance to thermal fatigue because the white layer is brittle and often contains cracks. The soft overtempered underlament exhibits poor resistance to crack initiation and propagation. The "white" layer can be machined away by mechanical means to restore thermal fatigue resistance to a limited degree, but it is physically impossible for some complex geometries.

Premium grade hardened die material (explained later in this chapter) accepts EDM without the undesirable effects described above. The resolidified white layer is soft but ductile and can either be hardened in service or by a simple aging treatment.

The schematic of the sequence of events that take place in the EDM process is presented in Fig. 2. In Box 1, an electrical field is created in the dielectric fluid between the work and the electrode. Box 2 illustrates an electrical spark between the work piece and the detail on the electrode. Box 3 describes how the heat from the electrical arc melts the material away from the work and releases it to the fluid. Note that some material is also removed from the electrode. In Box 4, the current is turned off when most of the melted material

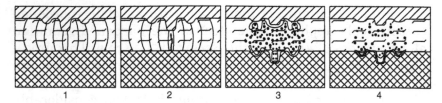

Figure 2

solidifies into spheres that are carried away by the dielectric fluid. However, some of the molten material solidifies on the surface of the work piece.

The EDM process is in wide use because of the economy in die casting die construction, but is controlled differently in tool shops where the economic benefit conflicts with the quality of the die cavity surface (Wallace and Schwam). The better shops reduce the amperage as the final die surface is approached to minimize the depth of the melted and solidified "white" layer, the untempered martinsite layer, and the amount of tempering. These layers are depicted in Fig. 3 to give perspective to the layers and hardnesses.

H-13 alloy is offered in regular and premium grades. Since the cost of the cavity die material is a very small portion of the total die cost, premium grade produced to meet NADCA standard No. 207 is recommended. The additional cost is not

	DEPTH	HARDNESS
MELTED & SOLIDIFIED	0.0003 - 0.0009 in.	56-58 [Rc]
	0.0009 - 0.015	50-54
UNTEMPERED MARTINSITE		
TEMPERED LAYER	0.0015 - 0.003	34 - 43
UNAFFECTED MATRIX		44 - 46

Figure 3

significant, so most die steel is produced under this standard with almost isotropic (independent of direction) structure. Homogeneous distribution of alloying elements when combined with fine segregation of carbides, sulfides, and oxides enhances toughness properties, primarily in the transverse direction.

It is essential that premium die material be used to construct, at a minimum, the 20% of the die casting dies that produce 80% of the castings. As premium grade material becomes more common, actual usage approaches more than half of all aluminum dies.

The premium grade acceptance criteria requirements cover:

- Chemical composition
- Hardness
- Microcleanliness:

  Sulfide
  Aluminate
  Silicate
  Globular oxides

- Ultrasonic quality
- Impact capability
- Shepherd grain size
- Annealed microstructure
- Microbanding designation level
- Response to heat treating

H-21 die steel is a tungsten-type hot work tool steel that contains approximately 9% tungsten and over 3% chromium. Other alloying elements are in the 0.3–0.4% range. This alloy retains some hardness even when cherry red and displays good wear resistance and toughness at high temperatures. These characteristics make it desirable for casting the highest temperature alloys.

Special die materials are used to control heat flow where thermal paths converge, as upon a core. These are tungsten-based alloys with high thermal conductivity. Anviloy 1150 has a very high resistance to thermal shock, but very little resistance to mechanical shock. Therefore, it is recommended that its size not exceed 4 in. in diameter.

With a thermal conductivity almost four times that of H-13, Anviloy 1150 could cause serious casting problems instead of enhancing castability, so it is suggested that a careful mathematical thermal analysis be made before inserting this material into an aluminum or magnesium die casting die.

A molybdenum-based steel called Mo-TZM has an even higher thermal conductivity which brings more risk so it is only available in small sizes. It is useful, however, for small core pins or inserts in the cavity.

Of course, only the die components that "see" the super heater liquid casting alloy are subjected to extreme thermal fluctuations. Thus, a partial list of recommended materials for die components is presented in Table 1.

Plunger tips are made both in H-13 steel and beryllium copper, the preference of die casters in North America because of its high thermal conductivity. Tips are almost always water

Table 1 Partial List of Recommended Materials for Die Composition

Die component	Material	Hardness
Cavity block—Zinc	P-20	300–325 BHN
Cavity block—Al, Mg	H-13	44–46 Rc
Cores	H-13	44–46
Core locks	SAE 6150	48–50
Core slide gibs	SAE 6150	50–52
Retainer block	SAE 4140	30–34
Ejector rails	SAE 1020	–
Ejector plates	SAE 1020	48–50 Case harden
Ejector pins	H-13	Case harden
Leader pin	SAE 1020	48–50 Rc Case
Leader pin bushing	SAE 1020	58–60 Case
Rack	SEA 6150	48–52 Rc
Pinion	SAE 6150	43–45
Runner block	H-13	44–46
Sprue spreader and bush	H-13	44–46
Shot sleeve	H-13	44–46
Wear plates	M-2	58–60

cooled because the biscuit is the most difficult element of the cold chamber to cool due to its high mass.

Hot chamber tips are made from H-13 steel and water cooled, but the tight fit is accomplished by a series of rings similar to piston rings in an internal combustion engine.

Heat treating is required with H-13 die materials to accomplish the working hardness of 44–46 on the Rockwell "C" scale, while P-20 is usually prehardened. H-13 die steel is very delicate as far as heat treatment is concerned: The time taken to drop H-13 steel from the austenitizing to the tempering temperature adversely affects the dimensional stability if it is too short, and the fracture toughness, if it is too long. This paradox means that proper heat treating is an awesome responsibility, not to be taken lightly!

With intermediate rates of cooling, carbides are ejected from austenite in different forms that offer different properties, especially toughness, even though the hardness does not change (Wallace et al.). The ideal metallurgical state is tempered martinsite with no carbons, which is almost achievable given modern technology for generating the cavity shape in the hard rather than the annealed condition. Martinsite is a distinctive magnetic needle-like structure that exists in hardened carbon tool steel as a transition stage in the transformation of austenite. It is the hardest constituent of the eutectoid composition.

Figure 4 provides a good guide for heat treating decisions that affect die life and dimensional stability.

It must also be understood that the purchaser of the castings also becomes the owner of the die since castings and tools

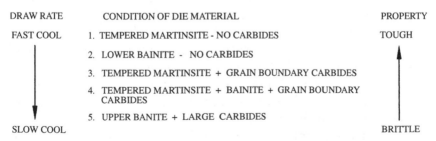

Figure 4

are normally bought as a package. Thus, a chain of financial responsibility develops all down the line. The die caster is responsible to the customer for the life of the die, and the tool maker is responsible to the die caster for the hardening and tempering, and the heat treater is responsible to the tool maker. It, therefore, behooves each link in the chain to request and carefully examine the heat treating records including furnace charts for discrepancies from specifications.

Now, to explain the above paragraph... The trick is in the cooling rate which is so closely limited that some metallurgists feel that H-13 is not heat treatable. It is important that the piece to be treated be raised to an austenitizing temperature of 1850°F ± 25°, and then lowered to a tempering temperature of 1300°F. Higher autenitizing temperatures up to 1975°F have been examined. The use of higher temperatures leads to excessive grain boundary growth, which drastically reduces the toughness because of grain boundary carbide precipitation. Thus, the piece goes through a transformation that is determined by the austenitizing temperature and cooling time.

The time/temperature transformation curve is commonly used by metallurgists to monitor this phenomenon. A typical chart is illustrated in Fig. 5. Four quenching rates are shown from fastest (more dimensional distortion) to slowest (lack of toughness) (DCRF, 1986).

Curve number 1 generates an ideal martensite structure, but the quench rate is too fast to be practical. Curve number 2 still results in martensite, but displays some grain boundary carbides and is only achievable with small tools that are oil or polymer quenched. Martensite plus bainite plus grain boundary carbides are created by curve number 3, which is a practical structure for medium to large tools. This is the slowest recommended quenching rate and is accomplished by a good gas quench in a fluid bed, etc. The structure represented by curve number 4 contains pearlite and lacks toughness even though tempered hardness may be correct. Center zone of large blocks will have this structure.

Basically, heat treatment for H-13 die steel requires that the work is placed into a furnace preheated to 600°F alone or

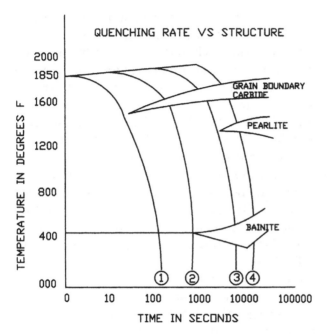

Figure 5

with other work of similar size and specification. The temperature is then raised to 1850°F, with steps at 900°, 1300°, 1550° to ensure uniformity, and then finally 1850°.

The work and furnace temperatures are allowed to equalize by soaking at each step for a minimum of 1/2 hr per inch of thickness at rates of 100–500°F, depending upon the complexity of the cavity shape. When the austenitizing temperature of $1800 \pm 25°$ is reached, the piece is soaked for 30 min for the first inch of thickness and 20 min for every other inch of thickness.

Drawing down the temperatures is called quenching, which is normally done in vacuum furnaces using inert gases. The cooling is delayed at 700°F to equalize the temperature before cooling further.

The desired metallurgical state for H-13 die material is martensite and it is critical to cool it so as to miss the nose of the ferrite curve or that will end up being the state of the steel.

After a temperature of 130°F is reached, the work is soaked for 2 hr at 1000°F per inch of thickness. Then it is dropped to room temperature when the first hardness check is made and monitored. Two more temperings are made by following the above procedure and heating to 1060° and 1000°.

All of this history (temperatures and time) is recorded on graphs that are called the furnace charts. Die casting dies cost hundreds of thousands of dollars and for the die caster or the customer to ignore these data is inexcusable.

Die life can be extended by surface treatment processes such as ion nitriding, ball peening, rocklinizing, soleveniting, etc. Ball peening is quite effective in peening the edges of fine cracks or checks together. Soleveniting continues to be very popular, and periodic stress relief on a regular basis will prove to be quite effective.

Preheat the die to a temperature above 300°F, but below the operating temperature to enhance fracture toughness, reduce the thermal shock of injecting superheated liquid metal into the die cavities, and to minimize the probability of gross cracking and die checking for longer die life.

This practice will also reduce start up scrap and increase machine up time. It should never be done after the die has been set in the casting machine because the machine will then be idle during the preheating. It is much more efficient to use a specially constructed angle iron frame that costs less than $500.00 rather than an expensive casting machine. However, very few die casting firms in North America use the special frame, and some do not even preheat to the proper temperature before starting a production run. The customer who actually owns the die would be well advised to monitor these procedures. Some of the methods used are listed below:

- Steam through the thermal lines
- Electric heaters between the die halves
- Gas torch in the shot hole
- Edge heating

Welding is not recommended but is acceptable in a few instances (Barton, 1963). When machining errors occur or unplanned modifications must be made while the die steel is in the annealed state (before heat treating), an arc weld (shielded metal arc) using H-13 coated welding rods is suggested.

TIG weld (gas tungsten arc) using a maraging or H-13 rod is suggested when welding hardened H-13 die material to repair shallow cracks or make small modifications.

For deep cracks or large modifications, an arc weld using coated austenitic stainless *steel* electrodes, topped with an H-13 or maraging steel rod using TIG welding, should be the procedure.

When a die is broken in two pieces, the pieces should be fastened together using arc welding with high-tensile-coated electrodes. The weld should be completed using an arc weld with coated H-13 electrodes.

Since this is a heroic strategy, several steps should be followed before and after welding. These procedures are outlined below:

- Degrease by cleaning with trichlorethylene or hot detergent
- Stress relieve by heating at 1000°F for 1 hr per inch of section plus 1 hr
- Then air or furnace cool
- Remove all cracks by grinding a "U"-shaped groove
- For deep cracks, remove enough metal to allow for at least two layers of filler weld and 1/8–1/4 in. of finish weld
- Clean the die by removing oxide, dirt, and discoloration, by vapor blasting, or by chemical cleaning
- Dry thoroughly
- Examine the die block for residual cracks by using crack detection methods such as die penetrant or magnetic particle inspection

When using coated H-13 electrodes:

- Keep rods clean and dry.
- Preheat the die block to at least 600°F, preferably 1000°F, using a temperature-controlled furnace. Never weld a die at room temperature.

- Deposit the weld bead using a low current value. Avoid making heavy deposits to minimize high stress levels.
- Peen the weld bead after each pass to remove slag and reduce stress. Do not allow the temperature to drop below 600°F.
- Examine the weld and cool the die away from drafts to about 100°F, let it set for 8 hr or it may crack.
- Stress relieve annealed dies by heating slowly to 1200–1300°F followed by air or furnace cool, then heat treat to desired hardness.
- Temper hardened dies at 50°F below the previous tempering temperature and alternatively temper between 1000°F and 1050°F for at least 2 hr.

For TIG welding using maraging steel filler rod:

- Clean and preheat to make sure the weld area is clean and free from grease, dirt, and oxide so that the weld will not be porous and cracked. Preheat between 300°F and 500°F and maintain this temperature.
- Weld with an adequate flow of argon-based gas; control the current for good penetration while avoiding undercutting. Use a TIG gun kept only for welding dies and use the correct grade of wire.
- Clean the weld after each pass and peen as necessary. Do not try to weld over porosity.
- Cool the die slowly and do not quench.
- Heat treat the welded die at 900°F for 4 hr to temper and stress relieve.

Soldering, described in an earlier chapter as a defect, is a complex physiochemical interaction between the die material and the superheated casting alloy (Shankar and Abelian, 1999). In this process, the cast part sticks to the die even after ejection. This phenomenon is a primary concern in aluminum die casting since the iron in the die material starts to dissolve in the aluminum casting alloy during the latent heat of fusion.

A series of intermetallic layers form at the die surface/casting alloy interface where the aluminum melt reacts with iron atoms. The intermetallic compounds have a lower heat transfer coefficient than the die material. Therefore, the melt in contact with these compounds cools at a slower rate and separates from the rest of the casting during ejection. The rate at which the compounds form is a function of the diffusion of iron species from the die material into the melt. Three or four layers form with spalling displayed at the top layer in raft-like intermetallic precipitates. The precipitates exhibit a surface tension effect on the aluminum that may contribute to soldering.

It is fairly well known that soldering occurs at hot spots (last place to solidify) on the die surface or where high gate velocity impinges upon a die detail. Thus, the soldering stages are erosion of the die surface by the superheated aluminum, corrosion and diffusion of the die material, and the accumulation of solder.

Since the incidence of soldering is strongly influenced by operating conditions that can be controlled, aluminum injection temperature, gate velocity, dwell time, etc. need to be within chart limits.

12

Mechanical Die Design

Mechanical design is one of the three engineering disciplines necessary for a practical and economical die design. While it is the most obvious, mechanical design has to be related to and based upon the other two previously discussed disciplines of fluid flow and thermal dynamics. This discipline is the most tangible, so most die casters focus on it first and, due to delivery time constraints, it usually carries the highest priority.

It is customary for the die casting tool engineers and quality assurance people to interact with the customer's product designers to transfer all available knowledge about the design and function of the shape to be die cast. Function, fit, cosmetic appearance, and assembly with mating components are the usual interests. Since the die casting process offers unique incentives to product design like fine dense grain structure and intricate detail, it is well worthwhile to intimately integrate the product with the process early on. Ultimate assembly can be simplified and total costs minimized.

For the best performance, however, the cavity pattern should first be oriented to allow the fewest tight bends in the runner and ample geography for effective gating and venting. This can be accomplished with a quick preliminary flow analysis that examines all of the possible gating options. Then, a quick thermal analysis will roughly locate the cooling channels and fountains to effectively manage the temperatures of both the casting alloy and the die components.

The mechanical design can then be wrapped around the two functional disciplines and the steels can be sized and other details like core slides established. This strategy is mentioned at the beginning of this important chapter because die casting dies are usually designed the other way around. Shoe horning is then required of the metal feed and thermal systems into the available space that is left after the die is designed to fit into the smallest possible die casting machine. This is typical, but not the best way to do the job.

Economy should be a major objective in mechanical design because, next to the die casting machine, the casting die is the most expensive tool involved in the manufacture of high pressure die castings. The design of this tool must therefore be compatible with the total quantity of parts required during the life of the part to be produced. Where the volume of parts is low, it is important to design an inexpensive tool. On the other hand, if the volume is very large, the design must focus upon productivity (cost and quality), long die life, and efficient maintenance.

Standard die components are commercially available that have much of the machining done, which save both time and cost. They include plates, die retainer assemblies, unit dies, and master die sets.

Plates includes cavity insets machined to appropriate tolerances. Complete die sets contain leader pins and bushings, ejector rails, sprue bushing and spreader pin, etc. Unit dies offer an economical advantage of multi-cavity operation from a single cavity die, once the unit retainer is purchased. However, unit dies tend to fall into the category of inexpensive tooling to satisfy low volume requirements.

Standardization must be developed by each die casting firm on an individual basis that serves their product and equipment mix the best. Such components as sprue bushings and spreaders, ejector pins, leader pins, guide pins for ejector plates, alignment guide blocks, screws, and dowels can be standardized quickly.

For example, ejector pins come in 32 different diameters and in several lengths so the die caster can limit the choice to four sizes and then cut the length to suit when needed. Thus, both new die construction and repair are simplified. This same strategy can be applied to other items like mounting clamps, shot sleeves, etc.

Cost justification is a function of both quantity and quality requirements.

The characteristics of a high performance die are:

- Calculated flow and thermal dynamics
- Quantified production strategies
- Premium grade die steels
- Sufficient material for strength and heat exchange
- Cavity details less vulnerable to mechanical and thermal stresses
- Optimum number of cavities
- Balanced locking force
- Efficient lubrication of wear surfaces

The advantages of high performance die design and construction are:

- Lower start up costs (first shot success)
- Less scrap—better yield
- Reduced die maintenance
- Longer die life
- Better casting quality
- Faster production rate
- Greater up time

The disadvantages are:

- Higher costs
- Requires modern skills and technology
- Longer delivery time

A graphic justification for high performance die design is illustrated by the graph in Fig. 1. In this typical case, the

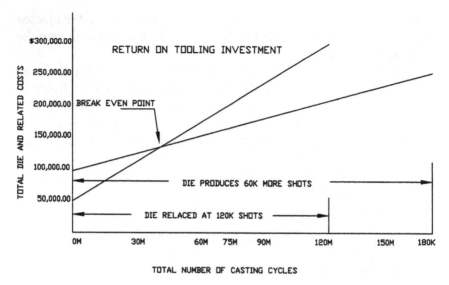

Figure 1

same part can be produced from a die costing between
$50,000.00 and $100,000.00 and each extreme is charted. It
can readily be observed that the break even point is at a
volume of about 45,000 shots. Beyond that lower quality,
higher maintenance, and less up time start to take their toll.

Most economic decisions on tooling relate to die life
because the buyer, who actually owns the tool (the die caster
has sold it to the customer), amortizes the initial cost of the
tool over the estimated working life of the die, which, in this
case, is 180,000 shots. The economic impact of superior performance from the die is clear.

Unit dies offer opportunities for economy in the absence of
volume. With a unit die, a different shape can be cast in each
station, in either single or multiple arrangements. Thus, in
cases, where usage does not warrant the tool cost of multiple
cavities, which would be too expensive to amortize over the life
of the shape, a single cavity is the choice. By running with
other casting shapes in the balance of the die, an equivalent
piece cost can be realized without the matching tool cost.

Normally, standard unit die assemblies that are commercially available are used in the trade. This tooling option is

configured in single-, double-, or four-station arrangements for either cold or hot chamber machines. When utilized by a custom die caster, castings for more than one customer can be die cast at the same time. There is nothing unethical about this practice because the die caster has invested in the unit holder and the customers have only purchased their particular unit station.

Core pulls are possible on two or three sides, depending upon the number of separate stations. When utilizing unit dies, one must be cautious to produce similar shapes together. Similarly, the characteristics of volume, surface area, and complexity must be met.

Die configurations also have a profound effect upon economics. Of course, as the number of cavities increases, the piece cost decreases. Production requirements always have a way of strongly suggesting the number of cavities. Many times, especially in the case of mating parts, a family or combination die configuration works well. In this case, multiple similar shapes can be produced in the same die, as long as quantity requirements are identical.

Identify the casting machine size and capacity to supply metal before any layout work is initiated, because the die must combine with the machine and cold or hot chamber shot sleeve to complete the work cell to produce the part to be tooled. It is important that the machine shot system can pump the required volume of metal to the die. With die casting machines that have been in service for several years with no preventive maintenance, the fast shot plunger velocity is as critical as the space between tie bars. To prepare for this, the metal feed strategy must be all worked out, which determines the number of cavities and locates the cavity pattern with relation to the shot center. This calculation is too often overlooked, which leads to shot end compromises that detract from die performance.

An important influence of the casting machine configuration on the die layout and cavity orientation is the location of the shot center. Usually, shot ends can be adjusted to three shot hole positions in the cover platen. They are center, 6 in. below, and 12 in. below center. For larger machines, these dimensions increase proportionately. Most hot chamber dies are center shot and all cold chamber dies should be shot below center.

Of course, the projected area multiplied by the optimum accumulator pressure (from the PQ squared diagram) determines the locking force required, which in turn determines the choice of available casting machines that are capable of producing the subject part. It is important to design the die for the least efficient machine if there are several of the same size available. Then, when the die is scheduled into the production plan, the planner will have a wide choice of machines and more flexibility.

The machine locking force required for a particular die is explained in Chapter 3 on the casting machine: The center of inertia of the cavity pattern must be at the center of the tie bar pattern if the locking strain on each tie bar is to be balanced. Die casting is quite forgiving, though, if this orientation is not perfect; the safety factor is usually great enough to overcome a small imbalance of locking force distributed over the four tie bars. It should be noted, however, that even though die casting machines are size rated according to the available locking force, it is normally the geography between the tie bars that limits the maximum physical die size the machine can handle.

Mechanical die casting die designs include a plan view of the ejector die half as viewed from the cover die position and a similar view of the cover die half as seen from the ejector die position. Sections are drawn as viewed from the operator's side. Other views are also common.

At least one full section should be cut through the shot center and a typical cavity with the die halves closed. Typical layouts and sections are included in Figs. 2 through 5 for reference.

Some of the details in the die sections of the figures are not necessarily in the location that relate to the plan views; they are shown only to illustrate details like push backs, support pillars, etc.

The example used here is a simple open and shut die intended to explain almost universal features that must be included in the mechanical die design. Threaded holes

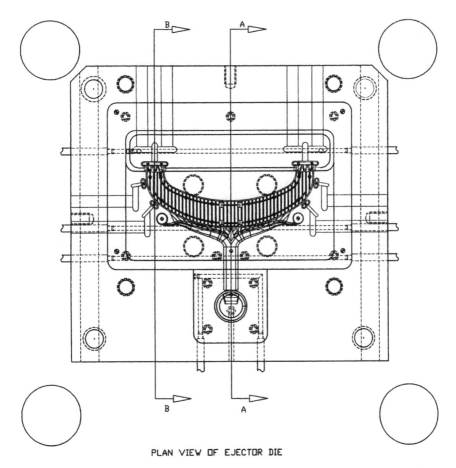

PLAN VIEW OF EJECTOR DIE

Figure 2

are strategically located on the top and both sides of each die half for efficient handling. The push back pins are larger than ejector pins to facilitate the return of the ejector plate to the closed position without straining them. The support pillars withstand the compressive force placed upon the ejector die during high pressure injection of the casting alloy and final intensification phase at the end of cavity fill.

When die size permits, it is desirable to show the plan view of the ejector die and the major section on the same

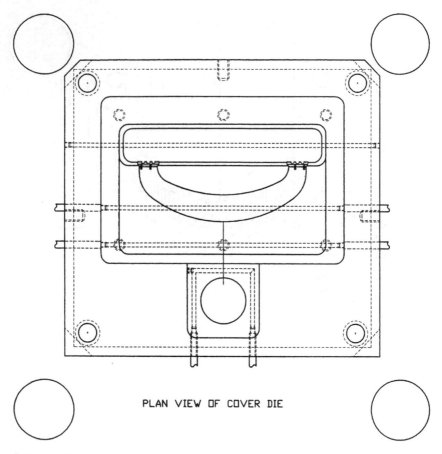

PLAN VIEW OF COVER DIE

Figure 3

sheet. The first sheet may also include the bill of material, change column, reference data such as machine information, and general notes.

Die details that are expected to require frequent replacement should be segregated on separate sheets or CAD files for convenience during production.

Detailed designs are sometimes prepared in which every single component is drawn and dimensioned. A complete design like this is more expensive and requires more time, and time is always at a premium. However, even though engineering time and cost are higher, a detailed design makes

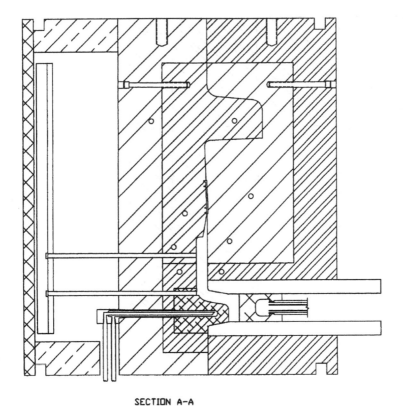

SECTION A-A

Figure 4

it possible for more tool makers to work on the die at the same time. Also, with total detail available, less experienced tool makers or apprentices can be involved, which more than off-sets the original investment in cost and time in engineering.

Dimensions necessary for the assembly drawing of the die include (Herman, 1979):

- Height
- Shut height
- Width
- Opening stroke
- Thickness
- Stock list

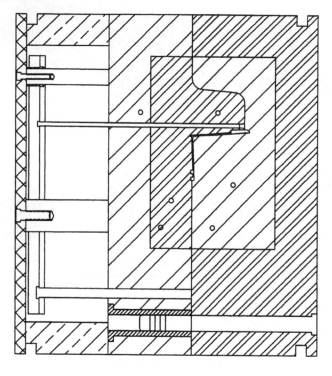

SECTION B-B

Figure 5

- Travel of all moving parts (ejector plate and core slides)
  The bill of material should include:
  - All major details
  - All standard purchased components:
       Ejector pins
       Leader pins
       Sprue spreaders
       Bushings
       Retainer blocks
       Unit die master
- Nominal sizes for all catalog items
- Finish sizes of die materials
- The quantity required of each detail
- Heat treat requirement

It is not necessary to describe small screws and dowels in the stock list.

General notes are suggested here as a guide for inclusion in the die design.

For cavity dimensions, work to the latest part print, applying high tolerance limits on coring and internal dimensions, and low limits on external dimensions.

Parting line (die blow), shrinkage, and angularity tolerances must be deducted from the product tolerance allowance to establish "tool tolerances."

Usually a 3D solid model of the product to be cast is used to define CNC tool paths for machining the cavity shape into the die steels. The base model must be dimensionally adjusted in accordance with the previous paragraph.

Water lines must not leak and must be tested accordingly. Stamp identification in location is easily visible in operating position.

CMM conformance to cavity dimensions or model and epoxy resin tryout shots of completed die shall be submitted for approval before the die is shipped.

Parting surfaces of the shut off must be uniformly spotted together until 90% of spotting dye is transferred to the opposite die half.

Shut height of die shall be flat and parallel within 0.005 in. TIR.

Tolerances on dimensions that are typical should be considered as noted here (in the event that a drawing is the genesis of the cavity shape rather than a computer model), with the notation "unless otherwise specified."

Two-place decimals ± 0.03 in.
Three-place decimals ± 0.5 in.
Angular tolerance in cavity $\pm 0°$ 15 min
Angular tolerance $\pm 1°$

Special gaging should be planned prior to detailing the cavity design because it quickly defines the casting details and tolerances that are critical to the product designer. The die dimensions should allow the tool maker no more than 10% of the tolerance specified on the part drawing or CAD file.

It is important to understand that cavity dimensions, alignment, etc. are machined into the die steels at room temperature, but that the die must produce the dimensions at a highly elevated temperature. The dimensional tolerances are specified for the purpose of accommodating the die casting process.

Parting line geometry is developed by a careful study of the shape of the part to be cast. This configuration dictates and defines any deviation from a flat parting plane which is called a parting line step. It also describes any holes, depressions, or undercuts that require cores that must move in a direction other than parallel to the die opening direction.

This geometry defines any mechanical restrictions to possible gate and vent locations, as well as other die details like overflows, alignment guides, false ejection, etc. that are external to the cavity.

In the case of parting geometry that is not flat, it is probable that the runner system or overflows will be located close to the edge of a parting line step. There is a danger of die steel that is structurally too thin between the runner and the step. Therefore, a minimum distance of 1/4 in. should be followed. A die section that illustrates this principle is shown in Fig. 6.

The die layout consists of a plan view of each die half and records the decisions that have already been made in the flow and thermal disciplines, and establishes those that are developed in this design phase.

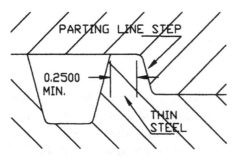

Figure 6

The die layout includes:

- Ejector/cover relationship to the casting net shape
- Cavity orientation
- Shot center relation to cavity (relates center of cavity cluster to center of tie bar pattern)
- Cavity insert size
- Shut off allowance
- Retainer size

The relationship of the ejector and cover components to the casting shape describes how the casting cavity will fit into each die half. Sometimes, the shape of the casting dictates the positioning of the cavity within the die halves. Many times, though, there are several options. Usually, external surfaces are located in the cover die because they will more easily pull away after solidification. Internal details are intentionally located in the ejector die since volumetric shrinkage during solidification is toward the inside. Figure 7 shows the four basic choices for locating the parting line for a simple shape. The cavity may be positioned on edge as in option 1; located entirely within either die half, as in option 2 or 4; or laid flat with some of the cavity in each die half as in option 3.

The final decision will be a function of convenient die construction, location of the metal feed system, position of flash planes for trimming, the relationship of critical dimensions to a datum point or plane, coring, ejection, shot removal, etc.

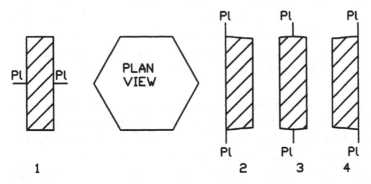

Figure 7

The thickness of each cavity insert is established by the distance of the extremities of the cavity in each direction from the parting plane. For both strength and heat conductivity, as a rule of thumb, another 2–3 in. is recommended, depending on size, to the thickness of the cover insert. Since the core stands up into the cavity from the ejector insert, an additional thickness of half that of the cover die will usually suffice unless the die sections are similar or even symmetrical. However, the strength and stability issue is discussed more quantitatively later in this chapter.

Cavity orientation is related to the fill pattern strategy and balanced feed system objective. However, once the orientation is determined, the cavity insert layout can be created.

The location of the shot center and its relationship to the cavity pattern location are also determined by the fill strategy, but this location is the basis for the shot position in the die casting machine for cold chamber configurations in which there is a choice of center and below center of the stationary platen. The shot block then must be designed around the shot sleeve or sprue post.

The center of the tie bat pattern should, where possible, be the central datum for the die layouts.

Cavity insert size can now be determined and then the retainer can be literally wrapped around it.

Shut off allowance is size sensitive, but a typical distance between the nearest cavity edge and the edge of the cavity insert of 2 in. is a good rule of thumb. The shut off distance around the shot sleeve or sprue post should be another 2 in.

The retainer size works best when another shut off distance of 4 in. between the insert edge and the outside edge of the retainer is established. Of course, the dimensions developed in this manner can then be adjusted upward to fit a standard die block.

The structural function of the design must address both the clamping force of the machine and the pressure applied to the liquid casting alloy during cavity fill. Both exert forces that can bend or distort individual die components.

The machine clamping force works to compress the die steels that fortunately exhibit extremely high compressive

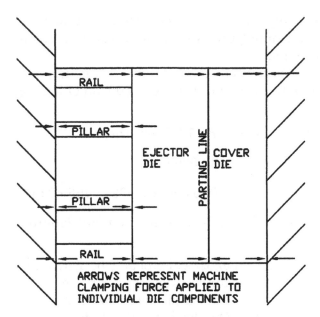

RAIL

PILLAR

EJECTOR DIE

PARTING LINE

COVER DIE

PILLAR

RAIL

ARROWS REPRESENT MACHINE CLAMPING FORCE APPLIED TO INDIVIDUAL DIE COMPONENTS

Figure 8

strength. This force is applied in the direction of the die closing and acts through the die retainers and in line with the ejector rails which represent the smallest compressive area of the die. Figure 8 illustrates these forces, which spread through the die retainers and ejector rails. Since the area of the rails is much smaller, the maximum compressive deflection will occur here.

Therefore, if the deflection in the rails can be minimized, the deflection in the die retainers can usually be ignored. The support pillars, which will be discussed later, are designed to resist the pressure on the metal and cannot be expected to resist the machine clamping force.

The amount of deflection in the rails can be calculated by a formula that states:

$$C = H \times 2000 \; F/M \times R_a$$

where C = total compression; H = height of rails; M = modulous of elasticity (use 30,000,000 psi); F = machine clamping force; R_a = total area of all rails.

The rail deflection should be below 0.002 in. When the height has been determined, the area of the rails can be calculated with this formula.

An example calculation is based on a die that is designed to run in a 1200-ton machine but requires 1000 ton of locking force.

If $C = 0.002$ in., $H = 14$ in., $F=1000$ ton, then

$0.002 = 14 \times 2000 \times 1000/30,000,000 \times R_\mathrm{a}$

$R_\mathrm{a} = 14 \times 2000 \times 1000/30,000,000 \times 0.002 = 466.67\,\mathrm{in.}^2$

If a rail thickness of 4 in. is chosen, then rail length = $466.67/4 = 117$ in.

On a die of this size, this length will probably be distributed around all four sides of the ejector retainer; thus, 30 in. per side.

The ejector rails, even though properly sized, must be evenly spaced to distribute the load evenly. Otherwise, if one spot deflects more than others, the nearest tie bar will not stretch enough and the locking force will be reduced and excessive flashing will result.

Support pillars, depicted in Fig. 8, are necessary to support the span of the ejector die over the area of the ejector system. They are usually round, but can be of any shape that will fit between inserted cores and ejector pins. The rails securely support the sides, but the machine locking force required to oppose the pressure applied to the metal during cavity fill puts a uniform load on this area.

Clearance holes in the ejector plate allow the support pillars that are attached to the back side of the ejector die to be supported by the moving machine platen surface. They are preloaded by designing them 0.004 in. longer than the rails that determine the space between the platen and the back side of the die.

Since the rails are designed to resist the total locking force applied by the machine, it is assumed that the support pillars only need to resist one-half of the total force generated.

The example is extended to calculate the area required for the support pillars where

S_p = area of support pillars, then

$$0.004 = 14 \times 2000\,(1000/2)/30{,}000{,}000 \times S_p$$

$$S_p = 14 \times 2000 \times 500/30{,}000{,}000 \times 0.004 = 116\,\text{in.}^2$$

Thus, if 4 in. diameter pillars were chosen, 1116/12.56 (area of 4 in. diameter) = 9.23, or 10 pillars are needed.

Thermal crowning is another significant structural factor that must be addressed because resistance must be provided in the mechanical die design. Otherwise, unpredictable dimensional deviations will occur, in addition to undesirable operating and maintenance conditions. The amount of unrestrained crowning that can be expected identifies this natural condition, generated by the temperature gradient between opposite sides of the die component that is usually in the range of 200°F. After calculating the gradient as discussed in Chapter 9, the three charts illustrated in Figs. 9 through 11 can be used to quantify the crowning. Each curve represents a different die temperature gradient.

To utilize the graphs, locate the length of the die component on the horizontal axis project up to the curve that represents the thickness. The amount of unrestrained thermal

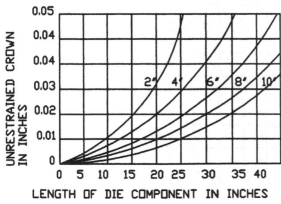

Figure 9

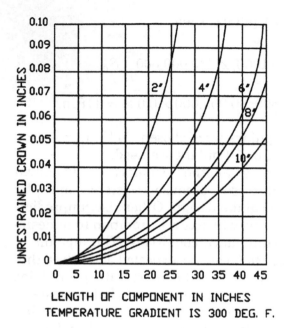

LENGTH OF COMPONENT IN INCHES
TEMPERATURE GRADIENT IS 300 DEG. F.

Figure 10

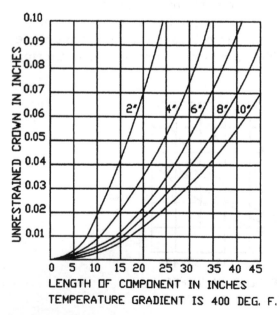

LENGTH OF COMPONENT IN INCHES
TEMPERATURE GRADIENT IS 400 DEG. F.

Figure 11

crowning is found at the vertical axis to the left of the inter-section on the appropriate curve.

The thermal gradients in a die casting die at operating temperature have a tendency to warp and bend the major die components. Figure 12 offers a schematic, albeit exag-gerated, visualization of this movement in hidden lines in an unrestrained cavity insert steel. In practice, screws are used to generate a force adequate to restrain thermal bending.

The force required to accomplish this can be calculated with the following equation:

$$F = 1{,}200{,}000{,}000 \times c \times L(T/W)^3$$

where F = force required to flatten crown in lb; c = expected unrestrained crown from Figs. 12 throught 14; L = length of component in inches; T = thickness of component in inches; W = width of component in inches.

An example calculation in which there is a cavity insert that is 8 in. thick by 15 in. long by 10 in. wide, with a tempera-ture gradient of 200°F. Then

$$F = 12{,}000{,}000 \times 0.004 \times 15 \, (8/10)^3 = 368{,}640 \, \text{lb}.$$

This is a tremendous force even though the expected crowning is only 0.004 in. Extremely large forces are necessary to over-come even this slight deflection. A 3/4 in. diameter machine

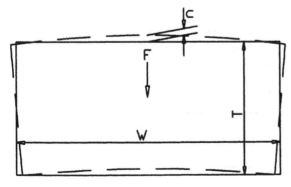

Figure 12

screw can be stressed to approximately 20,000 lb., and a 1 in. diameter screw will resist about 33,000 lb.

Then, in this example, $386,640/33,000 = 11.17$ rounded to 12 screws are needed to constrain the thermal crowning. Of course, one at each corner plus one in the center of the long side is the place to start in locating the screws. Now, a practical problem arises because the other six will be most effective near the center of the insert where cores or other details are usually clustered that conflict with the central strategy. Therefore, the final result is often less than perfect, and that is why some thermal bending is expected.

The dies must be aligned so that they will open and close in a direction that is exactly parallel to the shot line and machine closure. This is done in several ways, and the two methods usually used will covered here. Figures 13 and 14 graphically depict them.

Leader pins and bushings are normally employed to line up the two die halves with either the nozzle in the hot chamber, or the nose of the shot sleeve in the cold chamber process.

To preclude the die halves being put together backward, which could destroy cavity and core details if the machine were to close, one pin and bushing is offset from the other three. A convenient position for the offset is the top operator's side.

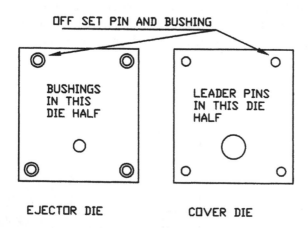

LEADER PINS & BUSHINGS

Figure 13

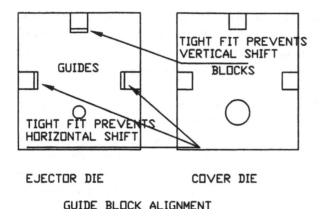

EJECTOR DIE   COVER DIE

GUIDE BLOCK ALIGNMENT

Figure 14

A less popular, but more precise, alignment is the use of three guide blocks. The center line of each block and mating ways are located on the horizontal and vertical die center lines.

Even with these alignment features, the die halves or other die components can be expected to shift with respect to each other, and provisions must be made for this phenomenon in the dimensional tolerance designed into the casting.

Construction tolerances and dimensional analysis define dimensional restraints that must be recognized because they affect the tolerances on the casting dimensions and the functioning of the moving parts of the die (NADCA, 1988).

There are many rules that determine construction strategy and only a few will be covered here to present some of the mechanical elements that must be addressed.

All casting dimensions and allowed tolerances have to be studied for dimensional variation that will describe the cavity and core detail.

Parting line separation is called die blow and occurs because the pressure applied during the injection of the liquid casting alloy exceeds 5000 psi and the die halves can be expected to blow apart by approximately 0.01 in.

A key factor in dimensioning cavity details is the allowance for shrinkage. All alloys experience a reduction in volume during the rapid solidification phase of the casting

operation. This is best explained graphically here. A nominal
10 in. component dimension is traced through the different
phases of the casting operation. In addition to volumetric
shrinkage, variations in die surface temperature exert a profound influence on size.

In practice, it is normally the responsibility of the die caster
to specify the shrink factor, which is the basis the tool maker
uses to revise every casting dimension accordingly. Each takes
a definite responsibility in the ultimate dimension of the as-cast
component. If the dimension of the die steel is within tolerance
(10% of part tolerance multiplied by the specified factor), the die
caster is responsible for correction if the casting is not to print. If
not within tolerance, it is the tool maker's problem. Figure 15
defines steel dimensions at different stages of the casting cycle.

Other construction tolerances cover the spacing between
cavity inserts and retainers, moving core slides, locking
wedge angles, etc.

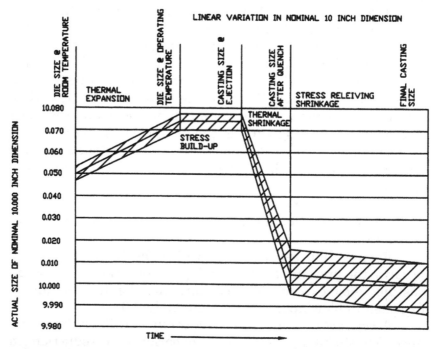

Figure 15

Draft angles, as with all casting processes, are a necessity since the casting alloy shrinks onto the male cores and away from the female cavity. Therefore, more draft is required on cores than on cavities.

Comfortable draft is 3° per side for aluminum and magnesium, but, with the right strategy, high quality castings can be produced with as little as $\frac{1}{2}°$ per side on outside surfaces. Zinc can be cast with half this draft.

Three-plate die design facilitates what is called center gating. This is a metal feed strategy that encourages a very diverse flow pattern with excellent venting opportunities. Sometimes, where the space is limited, a combination of a hot chamber style sprue is used to feed the center gate, but the cold chamber process is used to supply the liquid casting alloy. With the three plate die, the runner can even be configured to feed multiple gates or multiple cavities.

As described in the schematics shown in Fig. 16, it works by separating the cover die from the stationary plate when the machine opens, as illustrated in the middle sketch. The cover die stops moving at this point, and as the ejector die continues to move, the biscuit and runner system breaks off and drops away. The casting, including the sprue runner, stays in the ejector die and is then ejected in the usual manner as shown in the lower sequence.

Ejector systems operate mechanically when the ejector pins stop while the ejector or moving platen continues to move the cavity back and away from the casting. The ejector plate is drilled to accept the desired pattern of ejector pins. Ejector pins are commercially available in standard sizes with a head on one end. The purpose of the back plate is to contain the pins so the same pattern is drilled into it blind (not through) with hole sizes to accept the heads. A pattern of large diameter bumper pins is placed in the back platen of the casting machine to stop the movement of the ejector plate, while the dies continue to move apart. The length of the bumpers is designed to stop the ejector plate at the start of ejection.

As the ends of the ejector pins stop, so does the cast shot. As the ejector cavity continues to move away, the shot is

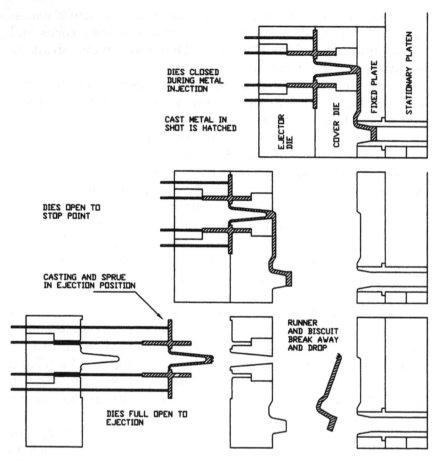

DIES CLOSED
DURING METAL
INJECTION

CAST METAL IN
SHOT IS HATCHED

EJECTOR DIE

COVER DIE

FIXED PLATE

STATIONARY PLATEN

DIES OPEN TO
STOP POINT

CASTING AND SPRUE
IN EJECTION POSITION

RUNNER
AND BISCUIT
BREAK AWAY
AND DROP

DIES FULL OPEN TO
EJECTION

Figure 16

cleared from the cavity so it can be extracted or dropped to a conveyor belt and transported to the trim operation. The schematic in Fig. 17 illustrates the typical ejector system.

The ejector plates are separated from the die retainer by rails which should be located top and bottom and on both sides to keep normal debris out of the ejector mechanism. These rails must be at least $2\frac{1}{2}$ in. wide. However, since they are pressed into the ejector platen, the mounting area is best calculated, as discussed earlier, to minimize the concentration of locking pressure upon this area.

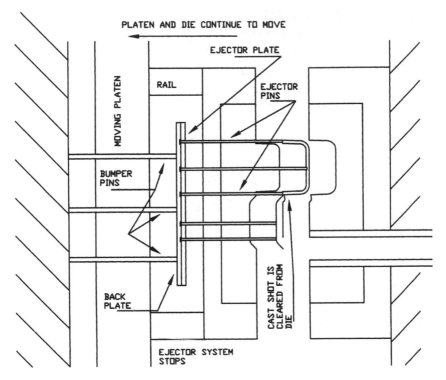

PLATEN AND DIE CONTINUE TO MOVE

EJECTOR PLATE

MOVING PLATEN

RAIL

EJECTOR PINS

BUMPER PINS

BACK PLATE

CAST SHOT IS CLEARED FROM DIE

EJECTOR SYSTEM STOPS

Figure 17

Movable details of the shape to be cast that present undercuts, reverse draft, etc. require special mechanical consideration. These configurations are formed with movable core slides, sometimes called core pulls. The two most common methods employed to move these slides are the mechanical angle pin and the hydraulic cylinder.

The angle pin method is illustrated in plan view in Fig. 18 and the next figure in section in Fig. 19. The angle pin provides the motion that is required, but it is important to note that the final positioning is accomplished by a wedge lock that takes its force from the locking pressure of the die casting machine.

The angle pin is fastened into the cover die retainer, and provides the basis and control of the core movement. The length of the pin and its angle determine the distance that

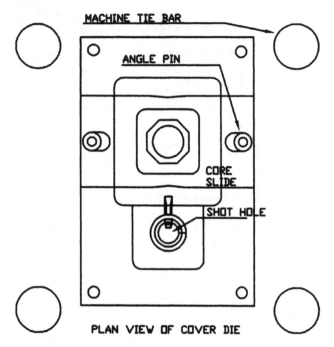

Figure 18

the core will travel. Normally, the angle pin becomes disen-
gaged when the dies are in the full open position. The spring
loaded detent shown in the next figure holds the core in the
"out" position when the dies are open and serves as the open stop.

   As the dies close, the leading end of the angle pin enters
the matching hole in the core block where the angle is the
same as that of the pin. The hold of the detent is overcome
and the core moves into its casting position.

   Clearance is provided between the pin and hole so that
the wedge lock can pull the core block away from the angle
pin to become the final locator. This clearance also prevents
binding between the two members. The angle of the wedge
lock is purposely designed 3–5° greater than that of the pin
for a tight lock that will resist the high metal pressure that
tries to force the core block away from the cavity.

   The pin angle can vary from 15° to 22.5°, with a maximum
of 25° for effective operation. An angle of 15° will move the core

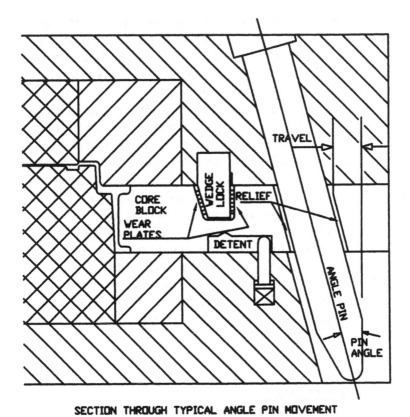

SECTION THROUGH TYPICAL ANGLE PIN MOVEMENT

Figure 19

approximately 0.27 in. per each in. of die opening, and the 22.5°
angle will move the core a distance of about 0.41 in. per in.

Angle pins are made from standard leader pins and the
conical end shape is the only modification necessary. Of
course, the head of the pin is machined flush to the bach surface of the cover die retainer.

Most die casters standardize the diameter and length of
angle pins for reasons of efficiency. When this is done, the
movement of the core slide is usually greater than necessary
but the additional travel is seldom detrimental.

The hydraulic cylinder is described in one form at
Fig. 20. Sometimes, it is possible to mount the cylinder directly
to the die retainer without the bracket. An advantage in this

design is that the movement can be independent from the open-ing or closing of the die casting machine. The wedge block prin-ciple can be used also for the final positioning and holding since it is not intended that the hydraulic cylinder does anything more than move the core slide in and out.

Consideration for processes after casting the near net shape is important because it is often possible to include sim-ple features into the casting die design that can assist second-ary operations. Such aids must be included in the mechanical design strategy early since as a design nears completion, there is a strong reluctance to incorporate changes.

Virtually all high pressure die castings require trimming of the thin flash that forms at the parting line during injection due to die blow. Therefore, this is the first item to consider. A die blow of 0.01 in. is normally expected, but the flash is thinner, maybe 0.003–0.005 in. The trim operation will only fold over such a thin flash that will still have to be removed with costly extra cleaning operations. A provision, what is referred to as a safety

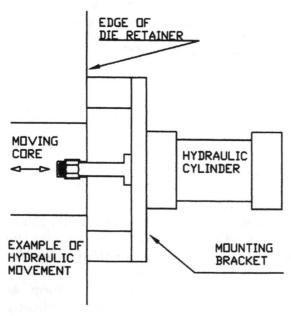

Figure 20

edge, forms a controlled flash line that is about 0.02 in. thick and $\frac{1}{2}$ in. wide. This trim bead or safety edge will trim cleanly.

If the safety edge continues around the total periphery, it can tie the overflows together to form a single piece of debris for more economical handling. It should be interrupted at the in gates, however so as not to disrupt the planned metal fill pattern.

Parting line geometry is a major factor in how well the shot locates in the trim die and also how well it is supported during the trim operation. Thin flash is really the easiest to trim, but when the parting geometry is in a plane other than parallel to the die opening, a different challenge is presented. One of many such conditions is described in the graphic in Fig. 21 that, when

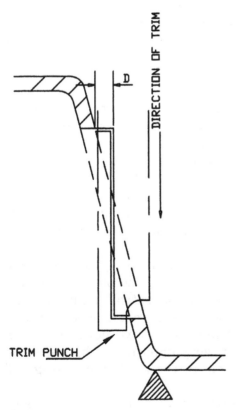

Figure 21

handled thoughtfully, can eliminate a secondary trim operation.

In this example, the vertical direction must be sheared rather than trimmed so this distance is limited to less than 1 in. to preclude additional deburring. The *D* dimension that will be directly trimmed is also important because it must be at least 0.02 in. greater than the allowance for a die shift of ±0.01 in. It is the draft angle that tolerates this strategy since adding to the bottom takes away from the top. Note that the trim edge cannot be directly supported.

Since it is necessary to have a gate into each casting and sometimes attach overflows, there are segments of the parting line that are thicker and will display "gate scars" after trimming. These scars can take several forms, as illustrated in the two drawings in Figs. 22 and 23. When the trim is too close as in no. 1, the edge is shaved and leaves a sharp burr. Another trim condition, shown in no. 2, is break out. No. 3 is an example of too loose a trim. The rounded lines represent how a polishing operation can fix these irregularities.

The trim line can be more uniformly controlled by an intentional plane of weakness designed into the gate and as described in the sketch in Fig. 24.

Rather than trying to shear the flash out of a cored hole, it is sometimes desirable to cast a rather thick slug across the hole surrounded by a weakness groove. Such a slug can be punched out from the back side with a round nosed punch so as not to unnecessarily skive the hole. The arrangement is illustrated in Fig. 25.

Locating holes or pins can be used to register the shot or casting either for the trim die or for a secondary machining operation. If they are not in a critical location with respect to the part design such as a metal saver, or are located external to the casting in an overflow, the *x-y-z* orientation to a clear datum is very important. This registration detail is an important item in the mechanical design of the casting die.

Cavity and run information is a feature that can easily be incorporated into each die cavity.

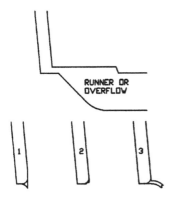

Figure 22

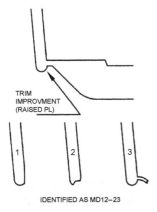

IDENTIFIED AS MD12–23

Figure 23

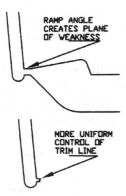

Figure 24

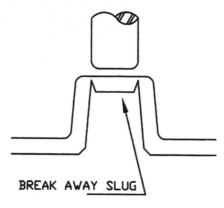

BREAK AWAY SLUG

Figure 25

Almost every part design makes an allowance for these data that includes:

• Part number and revision
• Die caster logo
• Cavity number
• Date stamp
• Die number

For economy, these data should be raised in the castings so that it may be depressed in the die. If this is not acceptable to the product designer, usually raised letters in a depressed surface will work. Sometimes different versions of the same part look very much alike, but are actually functionally different so the latest part data must be engraved or stamped into the die.

Many times, it is useful to know when a part was cast so it is helpful to include a date stamp in each cavity. This stamp usually includes the year and provides blanks where the month can easily be stamped with a simple punch mark. An example is displayed in Fig. 26.

Remember that the die design is an instrument to transfer detailed product information from the final assembly requirements of the user customer to the die shop and toolmakers. It is important to describe exactly what the die

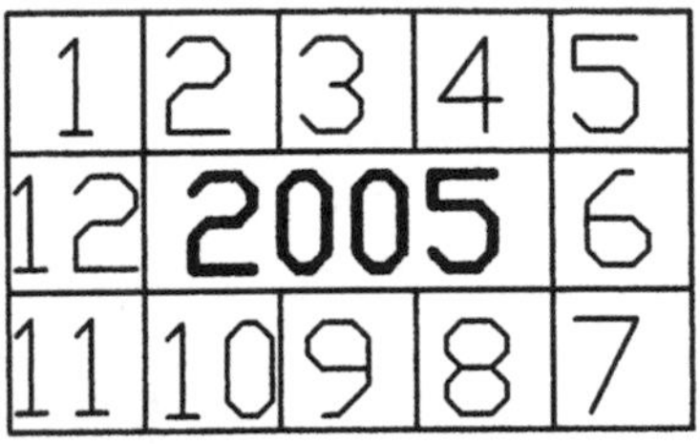

1	2	3	4	5
12	**2005**			6
11	10	9	8	7

Figure 26

should be when it is finished and ready for production. The design must be compatible with the machine and facilitate longevity, operation, manufacture, maintenance, identification, handling, and storage. In order to accomplish this, the design has to communicate these objectives clearly and concisely with adequate detail so that everyone involved with the project can understand.

Consider the mechanical design discipline as the package that contains strategies for metal and thermal flow, as well as forces applied by the injection of super heated liquid casting alloy and by the die casting machine.

13

Die Set Up Techniques

The most expensive equipment in a die casting plant is the die casting machine. Modern machines are magnificently efficient in that they embrace all of the technology that has evolved over many generations of machine design. Too many die casters squander this superb asset during die set up between production runs by taking too long to change from one die to another. The gap between acceptable set up time and actual is probably in the range of 1000%. Yes it takes 10 hr, even up to three working shifts or more, when it should take 1 hr! It does not make any sense to allow a million dollar die casting machine sit idle for the 9 hr difference because of a lack of organization.

Lean manufacturing technology is focused to attack this waste with a vengeance since one of its tenets is to completely eliminate waste from the value stream to set the end product apart from competition. Another form of waste is to produce an inventory of finished goods in excess of shipping requirements. It is indeed discouraging to walk through huge storage areas in many die casting plants. They are really designed to extend production runs far beyond schedules, and sometimes beyond purchase order releases, merely to support longer production runs between die changes.

Just in time delivery is another management discipline that expects shorter production runs and more die changes. The strategy here is to eliminate or at least minimize costly inventories of either in process or finished goods.

These are marvelous management concepts. However, the current gap between methodology and formal paperwork (in the form of Excel spread sheets, etc.) and actual manufacturing performance is far too great. Management generally appears to be too permissive in accepting previous production behavior and then wonders why costs are not competitive.

Management desire to survive in a free market is essential to eliminate the waste incurred from casting cells sitting idle between production runs while die are changed. In cases where efficient die set up is an improvement, such a radical change can only be accomplished by management with detailed comprehension of present conditions. It is important that top management be visibly enthusiastic on the factory floor, where foremen and employees can be stimulated to cooperate fully.

This cannot be accomplished by forcibly assigning such a difficult task to workers without properly educating them. Their willingness to respond to the demand will surely diminish as soon as they learn that the special efforts can hardly meet the objective of quick and multiple die changes per day and drastically reduce production run quantities.

Set up techniques vary from die casting plant to die casting plant. This procedure of removing the die from the machine at the end of a production run and replacing it with the next scheduled die is the cause of a disproportionate amount of down time. The focus of manufacturing is primarily on through put of product, so a higher priority needs to be placed upon die changing and inventory reduction. This change of emphasis may seem a paradox, but the serious blocking factor of die change between runs will eventually enhance productivity and return on assets exponentially!

The present condition is that too many die casting management teams *expect* it to take from one to three working shifts because:

- The required tools (wrenches, hoists, pry bars, etc.) must be gathered.
- Disunity of items needed, and lack of uniformity of screws, bolts, bumper pins, etc.
- The die must be located and transported to the machine.
- The shot sleeve, tip, or nozzle must be located.
- Mounting clamps and required gear must be checked out of tool crib and brought to the machine.
- Difficulty in wiring of hydraulic core slides.
- Excessive time to fit plunger tip to shot sleeve.
- Dirty work place.
- Electrical and hydraulic adjustments must be made.
- Cooling lines must be connected individually.
- Shot end and locking parameters must be set.
- Die must be preheated.

A casual observer who is not familiar with die casting would view the changing of dies as one of the most disorganized and inefficient events they have encountered. This is probably because it is viewed only as a necessary evil to be tolerated, so it is given a low priority as compared to the production of castings.

Have you ever waited for your car to be fixed at the dealer's service department where you have been advised that the work will cost $40.00–50.00 per hour? The first thing you see is that it takes 10 min to get it into place on the rack. Then the mechanic takes another 20 min to obtain replacement parts from the crib. After that it seems that he must discuss them with the service manager for another 15 min. Finally, the defective part is actually replaced in maybe 15 min. The bill for this "labor" is $45.00, but it only really required $11.25 worth of the mechanic's skills. Well, die set up in die casting is a lot like that.

Usually, only two set up people are used for all the tasks, which is why it takes so long.

Instead, why not follow a quick die change policy (Noguchi and Andresen, 1982)? To survive in the highly competitive world of die casting, waste has to be eliminated. Excessive inventories are a prime source of waste—production merely for the sake of long production runs is becoming a thing of the past. It is now common, especially in custom die casting operations, to see relatively short production runs and frequent die changes.

First, this requires commitment and dedication from top management to reduce nonproductive time. Remember, the average die casting machine represents a financial investment of about $500,000.00 and its time is valued at around $100.00 per hour, so it behooves an astute management to use it for production more than 80% of the time. This is what the bean counters call *return on assets*.

Standardization is necessary to make a quick die change a reality. A simple list is shown below:

- Carefully decide what has to be done.
- Classify each procedure into detailed tasks.
- Measure the time required to perform each task.

Divide all tasks into two groups.

1. Exterior group—Work that can be done while the casting machine is in production.
2. Interior group—Work that can be done only while the machine is down (not in production)

Reduce the number of interior tasks to an absolute minimum.

Write and distribute standard manuals that detail each entire exterior and interior procedure.

Balance the assigned times of interior tasks so that each set up person will take the same time as the others.

Perform only the work covered in the manuals ... avoid any other work that is not mentioned.

Now, does not this sound like working smart vs. working hard?

Exterior work has absolutely no effect on casting machine run time. It is, however, important that this work be done efficiently and without wasted time. It should include preparation of the following items, but not be limited to this list:

Cold chamber	Plunger tip and shot rod	Nozzle
Die clamps	Cooling pipes or manifolds	Eye bolts
Cooling hoses	Electric cord for core pulls	Goose neck
Wire rope or chain	Ladle bowl	
Plunger with rings	Bumper pins	

Organize necessary tools to:

Fasten die clamps	Assemble ejector rod	Assemble plunger tip to
Assemble core slides	Assemble hydraulics	shot rod

Still as exterior work, and after the above preparations, these tasks are necessary:

- Assemble movable core slides and related components onto the die, if possible. Install as many cooling hoses onto the die as possible, but do not disturb the present production run in the machine.
- Install eye bolts into all threaded holes on the dies by screwing them into the shoulder. Do not try to speed up the process by eliminating any of them.
- Transport dies close to the receiving machine (preferably onto the preheat rack).
- Organize all hand tools in order of their use in a suitable area near the receiving die casting machine.
- Add additional cooling hoses, if necessary.
- Place the cold chamber, plunger tip, and shot rod or nozzle and goose neck, next to the cover die.
- Locate the ejector rod next to ejector die component.
- Preheat the die halves.

Die preheating is essential to reasonable die life, but while it is commonly performed as interior set up work, it absolutely can and should be done as exterior work. It is a total waste of valuable casting machine time to delay the production of product until the die is heated up to operating temperature, which could take 2–3 hr. This task can easily be taken care of in space near the machine on an inexpensive angle iron rack rather than a million dollar casting machine.

The angle iron rack is illustrated in Fig. 1. There is no problem with setting a hot die into a machine because most dies are removed from the machine at just below operating temperature.

Three different preheating methods—propane torch, electric infrared heater, and hot oil, are depicted. Of course, only one approach is used on a single die, but it would be possible to use multiple means. These are the usual methods but they do not necessarily describe all those that are available.

Interior work should start only after all of the exterior work has been completed and, of course, after the previous production run in the casting machine is finished. The order of tasks for the interior procedure are listed here:

- Pull tie bar, if necessary.
- Attach the chain fall to the die halves and move them into position between the platens of the machine.

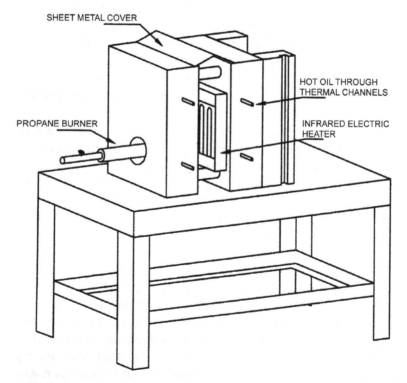

Figure 1

- Set the cold chamber into the stationary machine platen.
- Set the die in the proper location with respect to the cold chamber or sprue bushing, as the case may be.
- Set the plunger tip assembly into the cover die and connect the shot rod to the shot cylinder.
- Align dies both vertically and horizontally.
- Fasten the cover die onto the machine platen.
- Put the ejector rod in place.
- Fasten the ejector die onto the machine platen.
- Connect the hydraulic and electric lines to movable slides.
- Connect the cooling hoses.

Die set up time can be remarkably reduced and possibly cut in half, as long as these tasks can be done without interruption and this is the case, even without any special methodology or expensive installations or devices. If full commitment is made to quick die change, the improvement in up time of die casting machines and productivity can indeed be dramatic!

Efficient die set up technique has to be a function of die casting management that is dedicated to elimination of waste of available resources, i.e., time, money, or machines. Every company has its own culture that establishes the manufacturing environment, which includes working conditions, plant layout, customer orders and schedules, size and type of equipment, die designs, etc.

This concept introduces the following example of a procedure for one trained set up employee (die set up is this person's only job) and one skilled machine operator to set a die into a 250 ton cold chamber die casting machine. For simplicity, the machine is not automated and the same cold chamber is used. Note that two cranes are used.

Only interior work is described:

Operator's side of machine:	Helper' side of machine:
Undo die clamping nuts	Undo die clamping nuts
Remove clamps from die and cooling hoses	Remove clamps from die and cooling hoses

(Continued)

Operator's side of machine:	Helper' side of machine:
Place small bar in pouring hole of cold chamber to hold it in platen and hook chain fall to cover die	Hook chain fall to cover die
Operate machine to open dies	Lock out machine and slide out cover die from cold chamber
Remove cover die with chain fall and hook onto new cover die	Remove ejector die with chain fall and hook onto new ejector die
Replace bumper pins	Wipe fixed platen
Unlock machine and operate to establish new die height	Attach hydraulics to core slides
Set cover die and fasten clamp nuts	Set ejector die and fasten clamp nuts
Attach water hose to cover die and run water to measure flow rate	Attach water hoses to ejector die and run water to measure flow rate
Adjust tie bar tension	Adjust tie bar tension
Make first shot	Start next casting run

Accurate time allocation for each task is important so that one set up employee does not have to wait for the other.

Do not think of die casting as a stand alone operation. Almost any die can be automated with the technology available today, so it is not necessary for management to think of providing a worker for every casting machine. However, the nature of some of the equipment and adjustments is somewhat fragile, so it is advisable to have an operator nearby. Thus, another production operation is married to the die casting function to form a work cell.

Usually, an additional worker is needed for the trim operation—the worker runs the trim press, and tends the casting machine, robot, process settings, etc. If it is this easy to automate the casting operation, which is perceived to be the most complex, why not automate trimming and add a secondary operation? When the work cell starts to include secondary functions, the production volume has to be sufficient to warrant the necessary dedication of equipment to specific tools.

Much of the flexibility is lost when too much capital equipment is dedicated to specific projects.

The condition of tooling and equipment is critical to maintaining run time above the 80% level. Of course, the basic design has to be robust enough to withstand the rigors of an environment that is hostile to both people and machines, but this section is directed toward the procedures required to keep tooling and equipment in operational condition.

Too much time is consumed in repair work that is considered emergency and usually made while the die is in the machine, or, where the machine has failed, to patch it together well enough to finish the production run. A strategically planned preventive maintenance program will minimize these emergencies.

As a reminder, a strategic plan must deal with what must be done, what resources will be utilized, who will do it, and what will life be like when the plan has been accomplished.

Therefore, tooling must be regularly cleaned, and serviced, just like an automobile that is expected to perform efficiently during the expected span of its life. Of course, preventive tooling maintenance is done during nonproductive time to prevent the tool from breaking down.

Usually, a group composed of several disciplines (i.e., tool room, die casting, quality control, and engineering) conducts a review of the performance of a specific die at the end of the production run to determine what should be done, who will do it, etc. in writing. After the work has been completed, the same group compares the final condition to the planned preventive maintenance (PM), the die is then signed off to the production scheduling department, as available for the next production run.

The condition of the casting machine is a little more difficult to deal with because it is expected to operate all the time if return on assets is to be realized. However, a good planner has to accept the fact that each machine must be scheduled out of production for regular servicing if efficient operation is to continue...just like your car.

As with tooling, a similar group of different disciplines should regularly review the performance and repeatability of each die casting machine for experienced deficiencies and to see that regular maintenance procedures specified by the manufacturer are followed. The composition of this group should be similar to that for tooling except that the maintenance supervisor replaces the tool room foreman. Otherwise, die casting, quality control, and engineering (which includes process control) should be consulted.

A similar review of the conducted maintenance needs to be formalized before the machine is eligible to resume production.

There is a great temptation to keep the machine is production too long because of the pressures for production; when this is stretched to the point where the only maintenance is the emergency type, the results can be disastrous.

Standardization of dies is certainly desirable, but easier for captive die casters than for custom operations. The propensity for customers to move their dies from one custom die caster to another, makes for a large population of inherited dies with wide dimensional variations. However, this text would not be complete without suggesting some basic die dimensions to standardize. Figures 2 and 3 suggest standard

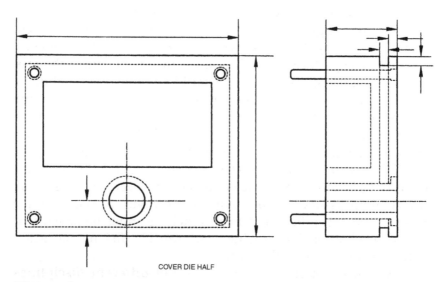

COVER DIE HALF

Figure 2

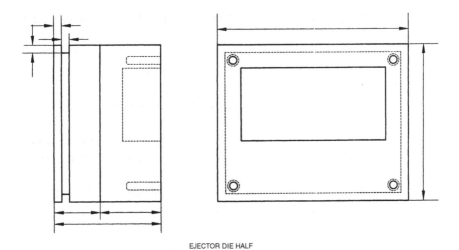

EJECTOR DIE HALF

Figure 3

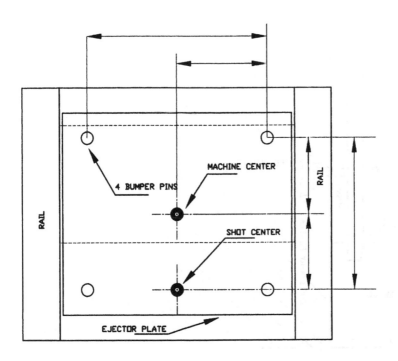

VIEW FROM BACK SIDE OF EJECTOR DIE

Figure 4

sizes for the ejector and cover die retainers for each size of die casting machine.

Figure 4 looks for uniformity in the ejection system that orients to both the machine and shot centers of a particular size of machine.

Consistent fits between clamps, tee slots on the platen, and grooves on the die retainer are described in Fig. 5. A spring can be incorporated for versatility and to reduce the number of separate parts when clamps are preassembled as exterior work.

One does not often see the nose of the cold chamber chamfered as illustrated in Fig. 6, but it makes the task of fitting the shot hole in the cover die onto it much quicker.

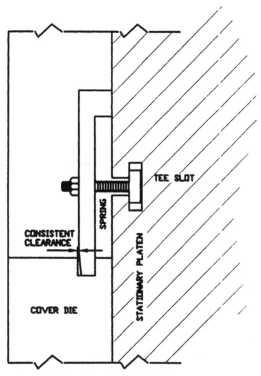

UNIFORM MOUNTING SLOTS AND AND CLAMP CLEARANCE

Figure 5

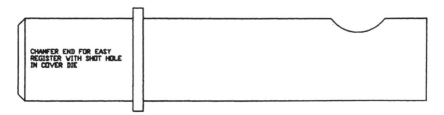

CHAMFER END FOR EASY
REGISTER WITH SHOT HOLE
IN COVER DIE

Figure 6

It has been suggested that certain patterns be established for internal cooling channels for standardization, but this is not included in this text since it limits options for efficient temperature management. Effective location of cooling channels is too closely related to the shape to be cast. Thus, it is the opinion of this author that connection of hoses not be standardized.

14

Die and Plunger Lubrication

Die and plunger lubricants position a cushion at the interface between the superheated casting alloy and the surface of the die steels. The insulating film is sprayed upon the die surface and prevents contact between it and liquid metal. The film must be strong enough to withstand turbulent metal flow. The lubricants used in die casting are often referred to as release agents; die lube is another term frequently used. High-pressure die castings would be impossible to remove from the permanent steel dies after solidification without the film of lubricant during ejection.

The science of rapid solidification that sets die casting apart from all other casting processes dictates that the casting alloy shrinks onto all male details of the net shape to be cast. In addition, all of the casting alloys, especially aluminum, have some affinity to amalgamate with the iron in the die steel.

The use of water-based lubricants have been almost universally motivated by safety, health, and environmental issues. A few solvent-based lubricants are used neat (without dilution) for die casting small zinc components. Zinc does not

have the affinity for iron that aluminum has and the temperatures are lower. Thus, the majority of die casting tonnage
utilizes water-based release agents.

The function of die lubricants contributed to the early
view of the die casting process as a black art. The composition
of materials used was mostly a trade secret and all any
one knew was that it worked. This lack of understanding
still prevails even though the industry has long ago out-grown
that reputation. This chapter will attempt to explain the
key role that lubrication plays in efficient high-quality die
casting.

An effective die lubricant imparts a thin invisible film
to the die surface which aids in the ejection of the solidified
casting from the die steels. Movable parts of the die are
also lubricated, which helps to minimize die wear. The film
applied to the cavity surfaces facilitates the flow of liquid
casting alloy during cavity fill since it tends to discourage
adhesion to the steel die surface. This mechanism is called
solder and increases with rising die surface temperature.
The selection of the correct release agent to prevent solder
is important.

The choice is best recommended by the lube supplier who
needs to be aware of the die surface temperature range from
low to high to intelligently suggest the most efficient product.
The chemistry of die lubricants has evolved to the point where
logical selection of additives, wetting agents, emulsifiers, and
polymers is beyond the average layman's understanding.

Actual performance trials are necessary to determine if
the product is compatible with the water analysis available
at the casting machine. A clean cast shape and die surface
are important under competitive cycle time conditions. Of
course, objectionable fumes or odors must be avoided.

Several things can happen to the lubricant during the
casting cycle that must be addressed when choosing the best
one for a particular die (Palidino, 1991).

- Elevated temperatures can break the lubricant down.
- High-velocity liquid metal streams, especially near
  the gate can wash the lubricant off the die surface.

- Sometimes a core detail will present a shadow during die spray so that a region of the die surface does not receive lubricant.

Materials that offer lower surface tension, better wetting, and binding characteristics help to offset the above conditions. Graphite or some families of boron compounds have lower surface tension and offer better creep characteristics, which allow the lubricant to spread and flow over the die surface more easily.

A general discussion of raw materials used in die casting lubricants is included here to provide a framework of basic knowledge for the die caster (Koch et al., 1989). Petroleum residue oils with a very high molecular weight residue wax form and retain films at higher temperature ranges. It reverts to a gelatinous mass at room temperature to suspend and retain dispersed pigments.

Animal and vegetable fats increase cohesion of residual films. Synthetic fats such as chemical esters, that are more polymer than petroleum oils and other fats, increase cohesion and wetting of metallic surfaces.

Pigments like graphite, aluminum, mica, and other powdered solids are indestructible at high temperatures and control viscosity and act as insulators.

Chemical additives are capable of changing the chemical composition of the die surface. It is structurally enhanced and cohesion is increased with oily ingredients whose antiwelding properties prevent oxidation or rusting of the die surface.

Special residual fluids are composed of organic compounds of huge molecular weight. They have critical temperature nodes at which viscosity is lost. During the extremely short cavity fill time (20–100 msec), a thick viscous layer with strong antiseparating properties is formed, which then reverts to a thin nonviscous layer.

Emulsifiers contain soaps, alcohol esters, and etholene oxide adducts. They are important to the formation of emulsion with otherwise immiscible materials. This contradicts the old addage that tells us oil and water do not mix. It explains how solvent-based release agents can be combined with water.

The whole mix is suspended in inexpensive fluidizers or carriers that remove heat through evaporation. This external heat transfer assists the internal cooling system in the die and has profound effect upon maintaining the desired die surface temperature.

Both water and solvent carriers dilute the concentrate and are the vehicle that applies the release agent, assisted by air, to the die surface. Carriers and diluents have the same function but are different in character and behavior.

Solvent carriers have generally been replaced by water carriers for reasons of good housekeeping, safety, and cleaner air. In the past, low flash-low molecular weight solvents like diesel fuel and kerosene were very effective at releasing the cast shape from the die cavity. May times a small explosion occurred on the die surface at every shot. Walking past die casting operations put one in mind of Dante's inferno. A major adverse effect was generated by the organic make-up that caused an undesirable carbon build up on the die surface.

The solvent carrier dilutes the other materials so that they are easier to apply in a thin even coat. It also acts as an insulator and evaporation media in the cooling process.

In theory, the first contact of the die spray with the hot die surface creates additional heat radiation. The low boiling point of this carrier causes it to immediately change to vapor, so that the die surface is cooled by evaporation. The other elements of the residue remain on the die surface and behave as an insulator and lubricant. The vaporization process continues until the boiling point of the highest constituent is reached. The highest rate of cooling takes place at this point and the deposit of the insulator is complete (Meister, H. R.).

Water as the diluent is much more common in that it minimizes air pollution and is safe since it has no flash point. If water is properly treated to remove minerals, no carbon or other deposits will be left on the die surface, but water cannot be considered chemically clean because of its mineral content.

Most die casters are aware of the problems caused during the die casting cycle by variations of mineral content in the water. Therefore, water treatment systems that also inhibit build up on internal water lines in the die are common in the industry.

The amount of minerals determines hardness which interferes with water-based lubricant mixtures. Split emulsions are unstable; in precipitation of solids in an immiscible form is possible; mineral deposits on the die surface can cause poor surface defects on the casting.

Water behaves differently than solvents upon contact with the die steels. It has only one boiling point at 212°F, and instantly converts to steam that removes heat as soon as it comes into contact with the hot die surface. After that, the lubricant is deposited on the die surface by extending the spray duration.

Internal cooling channels can be quickly clogged by water that contains large quantities of minerals such as calcium and magnesium salts, free iron, and sulfur. The hard water scale that forms can reduce the diameter so much that only a trickle flows through. Even when water is run through a softener, only part of the calcium and magnesium is filtered out.

If the same water supply is used for diluting external lubricant, the remaining minerals are suspended into a miscible state that will mix with the water and react the same way as those in untreated water upon contact with the die steels. Only deionization is effective. The deionization process removes all minerals, free iron, sulfur, and other impurities so that build up in cooling channels is impossible.

Mixtures are described as solutions when petroleum products are mixed with solvent carriers. When water-based concentrates are mixed with water, the mix is called an emulsion. There are different types of emulsions including molecular structures of water in oil, oil in water, semisynthetic, and synthetic.

Experience has shown that oil in water, in which oil droplets are surrounded by water, performs the best. Upon contact with the hot die surface, the body of water evaporates first and, in the process, heats the oil or concentrate. The oil is then deposited upon the die steel for release of the casting seconds later.

Water-based lubricants are normally used in the ratio of 30 parts of water to one part of lubricant for many aluminum castings. The other extreme of the range for small zinc shapes is 100 parts of water to one part of lubricant.

Application is usually done with a pressure nozzle supplied by either an individual reservoir at the casting site or from a central supply. Of course, a central supply dictates that the same mix and ratio be used on all shapes running at the time. The advantage is standardization but the disadvantage is that all casting shapes are not uniform or even similar so one size usually does not fit all.

Uniform and constant volumes of lubricant on each casting cycle improve metal flow and produce better looking castings. It is critical that byproducts of the lubricant not be encapsulated in the solidifying casting alloy since they will certainly become nuclei for porosity in the castings.

It is also important that a logical pattern be designed for each individual die that concentrates on regions such as long cores or deep ribs where sticking is expected. In other words, each spray nozzle must be aimed to be effective.

Control of the spray pattern and duration time is usually accomplished with automatic reciprocating sprayers that repeat the motion accurately each shot. Manual application is done with a single gun and sometimes is superior to automation because it is easier for human motion to reach difficult areas. Many times sprayers are attached to the extractor mechanism or the robot that removes the shot from the die.

Thermal properties characterize the lubricant in both the concentrated and diluted forms (Osborn and Brevick, 1997). *Surface tension* is a property that defines the molecular forces of the liquid to attract. It must be overcome to increase the surface area of each drop that typically occurs during increasing temperature.

The oils which have surface tensions lower than water are added to provide better wetting of the die surface. This relates to higher Leidenfrost temperatures that will be explained later in this chapter. The higher surface tension of water removes heat more effectively.

The contact angle is a factor of the surface tension and establishes the ability of the droplet of lubricant to spread upon the die surface. For example, mercury has a very high surface tension and a 25° contact angle. Water with a much lower surface tension has an angle of 110°. Most neat die lubricants are designed to have contact angles in the range of 160°.

Viscosity of the lubricant concentrate affects the mixing property and application. Proportional pumps on central mixing systems usually need to be recalibrated for different viscosities. Diluting the mix with water improves atomization at the spray head.

Thermal decomposition is examined by a thermogravimetric analysis in which the boiling point and vapor pressure determine how long the constituents of the lubricant are present.

Specific heat can be measured by a deferential canning calorimetry test which is not normally reported by die lube suppliers. It also has bearing on how long the lube is present during the die spray phase of the die casting process.

Thermal conductivity has been studied in lubricants used for the squeeze casting process and inhibit the heat transfer so that the interface between die steel and casting alloy is insulated. Such a lube design may not be good for high-pressure die casting.

The physical performance during application, cavity fill, and ejection is more understandable to die casters. Chemical reaction at the interface between the casting alloy and the die steel is a factor only if a physical change occurs. Thermal decomposition at operating temperature will detract from the efficiency of the lube. An important concern in zones of high heat concentration is resistance to soldering.

The wetting capability is probably the most important physical characteristic of die lubricant. The *Leidenfrost phenomenon* is a combination of surface tension and vapor pressure that effects and is often associated with wetting temperature (Alton et al. 1991). At temperatures above the Leidenfrost point, heat is transferred to the liquid lube through conduction in the vapor layer. At temperatures considerably above it, heat is transferred primarily by radiation from the die, which causes the spray droplets to vaporize and form a barrier. This prevents the droplet from wetting the die surface.

In this event, the die lubricant must cool down to the Leidenfrost temperature before it will wet the die surface and form a protective barrier. In Fig. 1 the cooling to the Leidenfrost point is expressed in curve *X–Y* and the wetting performance at curve *Y–Z* (Osborn and Brevick, 1997).

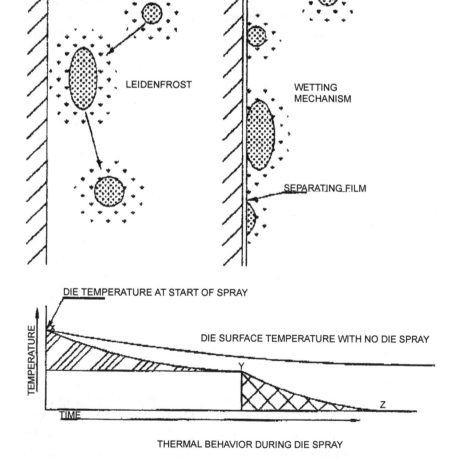

THERMAL BEHAVIOR DURING DIE SPRAY

Figure 1

For optimum management of die surface temperatures, the gradient between extreme highs and lows becomes a major objective. A carefully calculated internal thermal system design in the die casting die will minimize the gap. This is not the usual case, however, so some areas of the die surface will be above the Leidenfrost point and some below. This contradiction requires conflicting strategies that create confusing conditions and less than good yield and quality performance. It is best to maintain a more stable die surface temperature that will consistently react to different die lubricants that are designed for specific Leidenfrost temperatures. Figure 2 describes how two lubricants that display a range of temperatures perform differently. They are compared to shop water where evaporation times were normalized to the longest time.

A protective barrier is sometimes applied to the die steel to help prevent soldering especially when aluminum alloys are cast. At elevated temperatures aluminum can dissolve the iron so that the cast shape attempts to amalgamate with the die surface. Soldering occurs when the lubricant breaks down during cavity fill, in a hot region of the die surface, the liquid aluminum bonds to the steel die surface and attempts to dissolve the iron content. A material with desired properties is placed upon the die surface as a protective barrier.

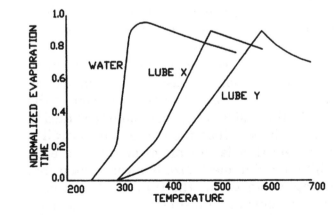

Figure 2

The mechanism for deposition can be by several different methods. Physical vapor deposition (PVD), chemical vapor deposition (CVD), physical chemical vapor deposition (PCVD), thermo-reactive deposition (TRD), and thermal spray are the usual methods used.

Suitable protective barriers require adequate adhesion to survive several thousand casting cycles, good mechanical properties (ductility, hardness, shear, tensile, and fatigue strength), corrosion resistance, high thermal conductivity, dimensional stability, compatible thermal properties with the substrate, and low surface wear.

Surface treatments like carburizing, nitriding, cold working, and work hardening are another method used. Here, the chemistry and/or microstructure are altered (Lewis, 2000).

Plunger lubricant is important because of the tight fit between the inside diameter of the shot sleeve and the outside diameter of the plunger tip. A small clearance of 0.001–0.002 in. per side is necessary to keep the superheated casting alloy from by-passing the plunger tip. The trick is that the movement must take place at elevated temperatures to expand the materials that slide together. In North America, some of the thermal imbalance is overcome by a beryllium copper alloy plunger tip, which has a much higher thermal conductivity than steel. The trade off is that it is softer and wears much more quickly than steel. Usually these tips must be changed every 10–15 thousand shots. For this reason, the rest of the world produces aluminum castings with an H-13 or H-11 steel plunger tip. Life of steel tips reaches toward 50,000 casting cycles.

Heavy graphited greases are used to lubricate and insulate plunger tips. Several methods are used that include dripping onto the back of the tip, brushing onto the front of the tip, and spraying into the inside of the cold chamber during retraction of the tip. It is best that the dosage be controlled by an automated system that is capable of monitoring each application. Graphite of a fine mesh size is incorporated into an oil base. The graphite insulates the tip and the oil base provides lubrication. To ensure both functions, usually more plunger lube is applied than is required because even slight misalignment will cause the tip to stick or chatter as it moves inside the cold chamber.

The housekeeping problem is generated when excess lubricant drips off the tip. The oil-based residue is very difficult to clean off the floor and machine surfaces. Water-based lubes have been used that leave a thin film of on the surfaces of the tip and sleeve. The residual mess can more easily be cleaned than when petroleum-based material is used. A tight fit and almost perfect alignment are required, however. Washable lubricants are also commercially available; they contain emulsifiers and contribute to a cleaner shot end environment.

A more desirable form of plunger lubrication has become more common today—a dry coarse powder free of fine dust. It is possible to maintain a clean shot end area since there is no oily dripping. It is applied into the pour hole of the cold chamber and is distributed via a burst of air (Camel and Munson, 1996).

It is designed to melt quickly at about 325°F so that it adheres to the surfaces of the chamber and tip. The partial melting is enough to provide a uniform coating. It does not melt and flow under further heating, and remains a smooth and very viscous liquid where it has to do the work. The physical form and the fact that the lubricant stays where it is applied make for a more efficient coating and performance. Liquid plunger lubricants flow toward the bottom of the cold chamber, which causes both chamber and tip to wear unevenly.

This is not as critical to the hot chamber die casting process used to cast zinc and magnesium alloys. Piston rings are incorporated with the tip, so a much looser fit can be tolerated.

15

Safety

The area around the die casting machine is hostile to humans because it is hot, dirty, at times smoky, wet, noisy, and dangerous! Die casting historically has not demonstrated an acceptable level of safety performance when compared to other metal casting industries.

Some type of protective clothing is required at every plant engaged in die casting. Usually, safety glasses are the minimum rule. At most plants, however, ear plugs to protect hearing, steel toed shoes for obvious reasons, and many times a helmet for head protection are also specified. In special areas like metal melting, protective clothing like heavy sleeves and spats prevent burns.

Speaking of noise, it is this writer's considered opinion, after many years spent around die casting operations, that the noise of the hydraulic pumps, constant spraying, electric motors, impact thumps, fans, trucks, sirens, horns, harsh public address systems, etc. forms a constant confusion of sounds that is so distracting that it dulls the other senses. This author believes that this contributes to poor communication and many of the mistakes that occur on the floor of any die casting plant.

ANSI/B152.1 is the safety standard that universally ensures that uniform safety procedures exist at all manufacturers engaged in high-pressure die casting. This American National Standards Institute document has bench-marked safety requirements for both equipment suppliers and die casters. It identifies and quantifies potential safety hazards associated with die casting machines and ancillary equipment (Mangold, 1997). 250 ANSI accredited organizations, from food processing to nuclear fuel handling, have produced consensus standards. The ANSI standards provide specific and concise instructions and considerations to the manufacturing community.

The Occupational Safety and Health Administration (OSHA) makes significant use of consensus standards as part of their review and audit of manufacturing facilities. The objective of OSHA audits is certainly consistent with that of die casting management, but there have been many disagreements due to the difference of priorities.

Metal handling is a major safety concern because of the superheated liquid state and very high temperatures involved. This subject is covered in detail in Chapter 5. The dangers to humans are the full range of burns and explosion directly from the casting alloy in the liquid state. It is difficult to make people understand that water on the surface of molten metal will merely boil off into steam, but water that gets below the surface expands into steam so rapidly that it explodes violently.

All metal handling equipment from the furnaces to the ladles to the cold chambers and goose necks is too hot to touch without a serious injury. Thus, extreme caution must be exercised by following all of the rules for behavior around molten metals.

Maintenance of automated equipment presents hazards that are exponential in nature when compared to the normal use of the same equipment in production. Lock down/lock out procedures are absolutely essential.

Signage requirements for capital equipment are governed by both ANSI standards and OSHA. The identification of potential hazards and warnings that are implicit to the

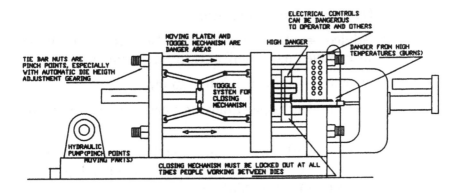

DANGER POINTS AT DIE CASTING MACHINE

Figure 1

operator or technician is required to be easy to read in understandable language appropriate to the population of employees in the plant. In some regions, Spanish is the language of choice. Figure 1 describes some of the potential hazards around a typical die casting machine.

Signage emphasizes the dangers and provides descriptive warnings. Some of the topics are outlined below:

1. Danger—high voltage

   Advice—before servicing, turn off, lock out/tag out main power disconnect. Do not modify electrical or hydraulic circuits unless authorized by manufacturer. Earth ground machine and electrical cabinet before turning on power.

   Warning—failure to comply can cause electrical shock, burns, or death.

2. Danger—high-speed moving parts.

   Advice—do not operate with gate guards removed or open. Do not reach around, under, over, or through gate guards while machine is in operation.

   Warning—can cause crushing injury or death.

3. Danger—high temperature.

   Advice—surface may be hot. Do not touch. Wear protective gear when working near this area.

   Warning—can cause burn injury.

4. Danger—high-speed moving parts between machine platens.

   Advice—with gate open, all safety devices must be on and functioning properly if entering area between machine platens.

   Warning—can cause crushing injury or death.

5. Danger—crushing injury at pour hole.

   Advice—keep hands and fingers out of pour hole. Do not place objects on bottom C frame shelf. See Manual for proper procedures on freeing stuck plungers.

   Warning—failure to follow safety procedures can cause crushing injury.

6. Danger—high-pressure accumulator.

   Advice—discharge all gas and hydraulic pressure before disconnecting or disassembling tank.

   Warning—can cause serious injury or death.

7. Danger—high-speed moving robot.

   Advice—interlocked perimeter guarding must be in place and functioning before operating robot.

   Warning—can cause serious injury.

Safeguarding devices are summarized here and their functions described.

Audible alarm—an electrical or mechanical signal, clearly discernable above the environmental noise to indicate a condition that requires attention.

Hard stop—rigid mechanical interface device that will prevent movement of a mechanical actuator past the point of contact.

Infrared sensor—an electrical device that monitors the light spectrum for infrared emissions. The sensor will send a signal to the controller when the infrared light is detected at or above a preset level.

Interlock switch—an electrical or mechanical means by which operation of a component is prevented or maintained unless required conditions are met.

Light curtain—a noncontact perimeter barrier device that will send a signal to the die casting machine controller if an object penetrates or interrupts the plane of the perimeter. This signal can then activate an alarm or interrupt the motion of the die casting machine to prevent damage or personal injury.

Motion detector—a sensor that monitors an object for motion. This is often accomplished with several infrared sensors in close proximity.

Physical barrier—a rigid boundary to prevent or deter access to a die casting work cell area. Cages, gates, walls, and fences are examples.

Safety mat—a flat pressure sensitive mat that is placed on the floor or platform around the casting machine, covering the full range of motion. If the mat is stepped upon, an alarm may sound or machine motion may be interrupted.

Safety mirror—a large wide angle placed in a strategic location to allow the operator to see otherwise obstructed views of the die casting work cell.

Ultrasonic sensor—an electronic transceiver that will send a signal to the casting machine controller if an object reaches a predetermined distance from the transceiver. It can activate an alarm or terminate machine motion.

Video monitor—video camera equipment located remotely from the operator that allows a clear view of an otherwise obstructed area, or an improved view of a visible area.

Visual alarm—a mechanical or electronic signal, clearly discernable above environmental distractions, that informs the operator of a situation that requires attention. It may be in the form of steady lights, flashing light, or a flag.

The plunger tip will seize in the cold chamber at times. It must be removed and refit before production can continue.

There are safe procedures for removing the stuck tip that need to be followed always, even if they are time consuming. There is a temptation to hold a bar or pipe between the open die halves and against the plunger tip. With the bar manually held in place, the ejector is closed against one end of the bar. The locking force of the machine then pushes the other end of the bar against the tip driving it back through the cold chamber until it moves freely. *This procedure is unsafe and must not be used* because pinch hazard is created between the dies. When the force of the machine frees the tip, the dies will close suddenly to cause serious injury.

A cheap and simple fixture is depicted in Fig. 2 that will eliminate the safety hazard. The alternatives are certainly less than desirable, but life and limb are at risk otherwise. For employee safety, the power to the machine can be shut off, with the platens open and the pressure accumulator isolated prior to working between the dies. The cover die and cold chamber can be removed so that the stuck tip may be removed on safe ground.

The most common cause is improper maintenance and operation. It is important ensure the diameters and fit of the shot sleeve and plunger tip, roundness, and alignment of the shot rod. Adequate lubrication is critical. The biscuit is the hottest portion of the shot at ejection so proper thermal

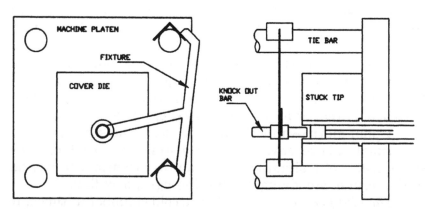

Figure 2

control as discussed in a previous chapter must be in place. In North America beryllium copper is used as the material for plunger tips because its thermal conductivity is greater than the H-13 steel material of the mating cold chamber. Thus, if excessive cooling is applied to the tip, it will shrink away from the shot sleeve and allow flash to build up and seize. If it runs too hot, it will expand too much and seize.

The biscuit will blow up if it is not sufficiently solidified after the plunger forces it out of the sleeve while still under high pressure. This occurs during the first few inches of opening when the biscuit bursts and spews out bits of liquid metal, hopefully onto the spit shields. However, if the machine operator or anyone else is in the way, they will be severely burned.

High voltage is a constant around a die casting machine, so caution must be observed at all times when working on electrical circuits. Die casting machine electrical systems are very complex and, apparently simple changes to a layout may result in extremely hazardous conditions when the machine is subsequently cycled. Therefore, only qualified and authorized technicians should work electrical issues. No changes should be made to the wiring without prior consultation with the machine builder.

High-pressure hoses are frequently used at various locations in a die casting work cell. They are used as hydraulic connections on the machine, or as flexible connections to core pull cylinders. They are used for flexibility and thus subjected to fatigue as they whip around. Original equipment hoses are supplied with the highest possible safety factor for their size, and, when worn, should be replaced with equal quality. A common application is to convey hot oil as a thermal medium to the die. Such hoses are armored to minimize leakage and wear. Severe burns result from being hit with 400°F oil!

The helper side of the casting machine is indeed a dangerous place to be during operation because all of the cycle controls are on the operator side and changes are frequently made that create a different motion pattern. Therefore, stand clear of this side during machine cycling. Laser beams are often used to automatically stop the motion when someone breaks the beam.

    Training is essential, given the hostility of the work place to humans. The die casting machine operator must be thoroughly familiar and comfortable with the machine. An intimate understanding of all its hazards is also critical. It is important to know the safe procedures and the use of these procedures must be policed and enforced rigorously.

Table 1   Check List

SAFETY PRACTICE	LIQUID METAL HANDLING	MACHINE OPEREATION	DIE SET UP	STOCK HANDLING	HOUSE KEEPING	TOOL ROOM
SAFETY SHOES	X	X	X	X	X	X
SAFETY GLOVES	X	X	X	X	X	X
SAFETY GLASSES	X	X	X	X	X	X
SAFETY HELMET	X	X?	X	X?	X?	
GLOVES	X	X	X	X	X	X
RESPIRATOR	X					
PREHEAT ITEMS TO BE IMMERSED UNTIL DRY	X	X	X			
NEVER EXCEED LOAD LIMIT	X	X	X	X	X	X
VENTILATING EQUIPMENT ON	X	X	X			
CHECK CONTROLS AND INTERLOCKS	X	X	X			
ADJUST TIE BAR TENSION		X	X			
KEEP CLEAR OF PARTING LINE		X	X			X
USE TONGS OR HOOK TO REMOVE SHOT		X	X			
LOCK OUT/TAG OUT HYDRAULICS AND ELECTRONICS		X	X			
REPORT OR REPAIR HYDRAULIC LEAKS		X	X			X
USE PROPER LENGTH EYE BOLTS			X	X		X
TURN EYE BOLTS TO SHOULDER			X			X
USE PROPER CHAINS			X	X		X
AVOID EXCESS SPREAD OF CHAINS			X	X		X
CHECK ALL SAFETY DEVICES	X	X	X	X	X	X
ORGANIZE LOOSE EQUIP.		X	X		X	X
WIPE OFF GREASE AND OIL	X	X	X		X	X
CLEAN FLOOR	X	X	X		X	X
REMOVE FLASH	X	X	X		X	X

Heat-related hazards are everywhere in the die casting plant. Open flames are so common, that it is important to know where flame is acceptable and where it is not. All die casting alloys, even though they are considered in the low-temperature category (melt below 2000°F), can cause third degree burns instantly because the metal is always in the liquid state during the production procedure. At least, liquid metal is visible, and the danger is obvious.

Occasionally, superheated liquid metal may escape at high velocity between the parting planes of the two die halves. This phenomenon is called die spit, and it can seriously burn any person who stands in line with the die parting plane. All employees and visitors to the die casting work cell must be cautioned accordingly. Power-operated gates that close as the machine platens close provide the most effective and common protection against die spiting.

Hot die surfaces are extremely dangerous, however, because the temperatures in the range of 400–500°F are invisible. Castings that have just been cast and not quenched present the same hazard. There is also no odor to the heat exposure, and it certainly may not be felt at the temperatures that exist; so the only way a person can be aware of it is to be trained or experienced in where to expect invisible heat.

Train die casting operators thoroughly to prevent accidents and promote safety. The check list in Table 1 is designed to focus the reader upon safe practices as related to the main die casting tasks.

It is impossible to eliminate the dangers from the production of high-pressure die castings. However, the risks to employees who work close to the heat and moving equipment can be reduced to a tolerable level by application of practical safety practices. It is a clear responsibility of management to take ownership of a comprehensive safety program that makes sense so that all exposed associates can enthusiastically buy into it.

References

ADCI Energy Bulletins (1976).

Altenpohl, D. (1981). Experiences with an energy action program.

Alton, T., Bishop, S. A., Miller, R. A., Y-Li Chu. (1991). A preliminary investigation on the cooling and lubrication of die casting dies by spraying.

Australian Die Casting Association. *Die Casting Bulletin* (Jan/Feb. 1995).

Baker, C. F. Processing molten magnesium in the die casting industry. NADCA congress paper G-T89–111.

Barton, H. K. (1963). How to vent die casting dies.

Camel, T. H., Munson, H. G. (1996). Powdered plunger lubricants for die casting. *Die Casting Engineer* Jan/Feb.

Chu, Y., Cheng, P., Shiupuri, R. Soldering phenomenon in aluminum die casting; possible causes and cures. The Ohio State University, NADCA Congress paper T93–124.

Crossen, H. M. (1972). Implementing technical change in a manufacturing organization. Beloit College.

CSIRO (1991). Computer aided thermal analysis for die casting.

CSIRO (Australia) (1992). Computer aided runner and gate design for pressure die castings.

DCRF (1986). Welding H-13 die casting dies.

Davis, A. J., Murray, M. T. (1981). SDCE congress paper, G-T81–123.

Die Casting Development Council. *Product Design for Die Casting.* 2nd ed. Lagrange, IL.

Doehler, H. H. (1951). *Die Castings.* New York: McGraw-Hill.

El-Mehalani, M., Miller, R. (1999). On manufacturing complexity of die-cast components. The Ohio State University.

Dorsch, S. (1991). Eliminating the negative effects of EDM through the strategic selection of die steels. Crucible Research Center.

George Group (2003). Achieve breakthrough results. Internet Nov.

B. Guthrie P. E. (1995). Effective furnace design for receiving molten aluminum delivery.

Hedenhag, J. C. (1989). Real time—closed loop system for shot end. Tymac, NADCA congress paper T89–022.

Herman, E. (1979b). Die cast dies: designing. SDCE 1979.

Jorstad, J. L. Applications of 390 Alloy, and Update. Reynolds Metals Co.

Jorstad, J. L. (1985). An overview of the need for melt cleanliness control.

Karni, Y. (1991). NADCA congress paper T-91-oc2-1991.

Koch, J. W. M., Trikfeldt, S. W. S., Koch, G. M. B. H. (1989). New Technology in melting aluminum for die casting industry.

Lee, I. S., Nguyen, T., Leigh, G. M. (1991). Cooling effect of lubricant sprays for die casting dies.

Lewis, J. (2000). Engineers try eight different coatings to find most effective and economical one. *Design News*, Dec.

Mangalick, M. C. (1976). Molten Metal Systems.

Mangold, V. L. Evolution of die casting safety. NADCA congress paper T97–112.

McClintic, R. P. Die cast process and tooling improvement. CDC, NADCA congress paper T95–096.

Meister, H. R. The functions and properties of die casting lubricants.

NADCA (1988). Heat treatment of H-13 die casting tool steels.

NADCA Product Standards (1994).

Neff, D. V. Principles of molton metal processing for improving die casting quality. NADCA congress paper T91–036.

Nicol, D. (2003). Lean methodology can restore competetivness for the domestic die casting industry. *Die Casting Engineer*, July.

Noguchi, T., Andresen, W. (1982). Quick die change guidelines.

NSM/OSU (1991). Comparison of methods for characterizing porosity in die castings.

Osborn, M. GMPT, General Motors Power Train, Brevick, J., OSU, The Ohio State University. Laboratory characterization of die lubricant performance.

Palidino, A. C. (1991). Die casting lubricant research meets die casters needs. *Die Casting Engineer*, May/June.

SDCE (1987). Die casting defects.

Shankar, S., Apelian, D. The role of aluminum alloy chemistry and die material on soldering. NADCA Congress paper T99–083.

Von Tachach, B. Some aspects of feed design for pressure die casting. SDCE congress paper G-T79–094.

Von Takach, B. V. (1996). ABC's of the PQ squared concept. *Die Casting Bulletin* (Australia).

Walkington, W. Die casting defects. (several photographs).

Wallace, J., Roberts, W., Hakulinen, E. Influence of cooling rate on the microstructure and toughness of premium grade H-13 die steels. NADCA Congress paper T99–102.

Wallace, J., Schwam, D. Control of die steels and processing to extend die life. NADCA Congress paper T99–102.

Index

9 780367 393564